George Bingham Fowler

Chemical and Microscopical Analysis of the Urine in Health and Disease

Second Edition

George Bingham Fowler

Chemical and Microscopical Analysis of the Urine in Health and Disease
Second Edition

ISBN/EAN: 9783744686167

Printed in Europe, USA, Canada, Australia, Japan

Cover: Foto ©berggeist007 / pixelio.de

More available books at **www.hansebooks.com**

CHEMICAL AND MICROSCOPICAL

ANALYSIS OF THE URINE

IN HEALTH AND DISEASE

DESIGNED FOR PHYSICIANS AND STUDENTS

BY

GEO. B. FOWLER, M.D.

EXAMINER IN PHYSIOLOGY, COLLEGE OF PHYSICIANS AND SURGEONS, NEW YORK, VISITING SURGEON TO THE NEW YORK DISPENSARY, FELLOW OF THE NEW YORK ACADEMY OF MEDICINE, MEMBER OF THE NEW YORK COUNTY MEDICAL SOCIETY, ETC.

Second Edition

REVISED AND ENLARGED, WITH EIGHTEEN ILLUSTRATIONS

NEW YORK
G. P. PUTNAM'S SONS
182 FIFTH AVENUE
1876.

COPYRIGHT.
G. P. PUTNAM'S SONS.
1876.

PREFACE.

In this, as in the first edition, the object has been to present the most practical and important features of the subject. The book has been carefully revised, and somewhat enlarged by the introduction of new matter, but the same general arrangement is preserved.

The decimal system of weights and measures has been adopted, also the new chemical nomenclature and notation. To meet the requirements of those not familiar with the decimal system, equivalents in grains or ounces are usually given in brackets, and a table of comparative values is inserted on the last page.

OCTOBER, 1876.

PART I.
1. CHARACTERS OF NORMAL URINE.
2. EFFECTS OF REAGENTS UPON NORMAL URINE.

PART II.
CHARACTERS OF ABNORMAL URINE.

PART III.
URINARY DEPOSITS.
1. THOSE WHICH ARE NATURAL CONSTITUENTS OF THE URINE, EITHER SEPARATELY OR IN COMBINATION.
2. THOSE WHICH ARE FOREIGN TO ITS COMPOSITION UNDER ANY FORM.

PART IV.
ACCIDENTAL INGREDIENTS WHICH DO NOT FORM DEPOSITS.

PART V.
QUANTITATIVE ANALYSIS.

PART VI.
CALCULI AND GRAVEL.

ANALYSIS OF THE URINE.

PART I.

Characters of Normal Urine.

The **Composition** of the urine per thousand parts is as follows:

	Water	950.00
ORGANIC.	Urea	26.20
	Creatinine	0.87
	Sodium and potassium urates	1.45
	Sodium and potassium hippurates	0.70
	Mucus and coloring matter	0.35
INORGANIC.	Sodium biphosphate	0.40
	Sodium and potassium phosphates	3.35
	Lime and magnesium phosphates	0.83
	Sodium and potassium chlorides	12.55
	Sodium and potassium sulphates	3.30
		1000.00

These proportions are not invariable, but as in all animal fluids, greatly depend upon diet, age, sex, occupation, etc.

A glance at the above table is sufficient to reveal

the fact that the urine is practically a watery solution of urea and inorganic salts, the chlorides being especially abundant.

UREA—CH_4N_2O.

This is the most important, and next to water, the most abundant ingredient of the urine. It is present in the urine of all animals in large proportion, except that of birds and scaly amphibians, where it is almost wholly replaced by the urates.

PHYSICAL AND CHEMICAL CHARACTERS.

Urea is a crystallizable substance, and has the chemical composition given above. The crystals are white, delicate needle-shaped prisms, very soluble in water and alcohol, but sparingly so in ether. It is neutral to test-paper, has a cooling taste like saltpetre, and is without odor. It never appears as a spontaneous deposit in urine. If kept protected from the atmosphere, or in a pure watery solution, urea does not decompose; if, however, it is boiled for a long time in pure water or a few moments with an alkali, or heated with water in a sealed tube above 100° C., a change is effected whereby *ammonium carbonate* is produced by the appropriation of the elements of two molecules of water, thus:

Urea.	Water.	Ammonium Carbonate.
CH_4N_2O +	H_4O_2 =	$(NH_4)_2CO_3$.

The same phenomenon occurs when a watery solution of urea is brought in contact with decomposing organic matter. Hence the ammoniacal odor of stale urine.

A peculiar interest is attached to urea as being the first organic product which was artificially produced. It was obtained by Wöhler by mixing the vapors of cyanic acid and ammonia:

$$\underset{\textit{Cyanic Acid.}}{CHNO} + \underset{\textit{Ammonia.}}{NH_3} = \underset{\textit{Urea.}}{CH_4N_2O}.$$

The first result of this process is the formation of a neutral cyanate; but on heating this is converted into urea. To-day urea is artificially obtained by several methods from cyanogen compounds, and by the action of oxydizing agents upon albuminous matters and upon uric acid. A method which yields it in great abundance is the following: Equal quantities, by weight, of potassium cyanate and ammonium sulphate are dissolved in a little water, and then evaporated to dryness over the water-bath. The residue is boiled with strong alcohol, which dissolves out the urea from the potassium sulphate and ammonium sulphate. The alcoholic solution is now filtered, concentrated, and allowed to cool, when the urea will show itself as a crystalline deposit.

To obtain urea from urine involves a much less satisfactory process, in that it requires more manipulation, and necessitates the employment of an immense quantity of urine to commence with, in order to secure an appreciable amount of the substance sought. The process is as follows: Urine is evaporated to a syrupy consistency, the deposit of phosphates and urates being filtered away as they appear, and about one-quarter of its volume of strong nitric acid added. A crystalline deposit of *urea nitrate* is immediately formed, which is made still more abundant by cooling. The whole is then to be thrown upon a filter and the crystals dried between folds of bibulous paper. They are then dissolved in a little warm water and the solution heated with an excess of barium carbonate.

The nitric acid quits the urea, setting it free, and unites with the barium to form barium nitrate; the liberated carbonic acid manifesting its escape by effervescence. A little concentration will cause the barium nitrate to crystallize, when we get rid of it by filtering. Now evaporate our watery solution of urea to dryness over the bath and extract with as small amount of alcohol as will dissolve the residue. On cooling the urea will crystallize out.

Urea behaves in some respects like an organic base. It unites readily with nitric and oxalic acids to form the nitrate and oxalate respectively. Both of these products may be obtained from the urine, it being only necessary to evaporate this fluid down to about one-fourth its volume, allow it to cool and add an excess of the acid. In the case of nitric acid there will be a very abundant precipitate of *urea nitrate* in the form of flat six-sided scales more or less colored by the urine pigment, and overlying each other to such an extent as to almost conceal their true form.

Urea has a much greater affinity for oxalic than for nitric acid, and if we add a strong solution of the former to the deposit already obtained the nitrate will be decomposed, and a copious fawn-colored mass of *urea-oxalate* crystals appear. The crystals of this latter formation are very similar to those of the nitrate, being rhombic plates and prisms. They may however be distinguished by their somewhat smaller size and the appearance among them of a few rather thick four-sided prisms.

PHYSIOLOGICAL AND PATHOLOGICAL RELATIONS.

Urea is the product of the retrograde metamorphosis of the nitrogenous constituents of the body. It is not formed in the kidneys, as once supposed, but exists in the blood of the general circulation in the proportion of about 0.16 parts per thousand. It has also been discovered in the lymph and chyle, the humors of the eye, and in the perspiration.

Whether urea is produced in the solid tissues or in the blood is not determined, but it is quite evident that the kidneys simply serve as eliminators of it; because it is found that the renal veins contain much less urea than do the renal arteries; and if the latter vessels be tied there is a very marked increase of urea in the blood.

Urea is excreted at the rate of about 35 grammes (525 grains) per day. But this process depends very greatly upon several conditions which are never to be overlooked. The most important are age, sex, bodily weight, food and muscular activity.

Women and children produce less urea than men, though in children there is more in proportion to the weight. (Harley. Scherer. Becquerel.) Large people excrete more urea than small ones. The proportion is about .5 per thousand parts of the weight. All observers agree that the excretion of urea is augmented under the influence of a nitrogenous diet, and diminished to a minimum by the use of food strictly non-nitrogenous.

Five-sixths of all the nitrogen ingested has been found in the urinary solids; and the entire quantity is to be detected in the urine and feces.[1] Practically urea represents all the nitrogen excreted.

Whether muscular exercise affects the amount of urea discharged has been long a mooted question and the subject of numerous careful and laborious experiments. Liebig promulgated the doctrine that the tissues consume their own substance with the result to eliminate an amount of nitrogen proportionate to their activity. This view, however, has lately been assailed and in turn supported by many eminent observers.

According to Parkes there is no increase of nitrogen—*i. e.*, urea—during unusual muscular exercise, but after such a period the increase is very striking.[2]

Fick and Wislicinus,[3] two German experimenters, estimated the amount of urea discharged by themselves before, during and after the ascent of a high mountain on foot, the diet consisting of non-nitrogenous food. The results obtained showed that there was less urea eliminated for a given time during the work than before. At night, having arrived at the top, they partook of a mixed meal which had the effect to augment

[1] Parkes. Proceedings of the Royal Society, June 20, 1867, and Lehmann, Physiological Chemistry. Am. ed., vol. 1, p. 151.

[2] Proceedings of the Royal Society, Jan., 1867. Vol. xv. Id. vol. xvi. Id. March, 1871.

[3] "On the origin of muscular Power," Philosophical Mag. (Supplement) vol. xxxi., 1866.

the nitrogenous discharge, but it did not with either individual equal that excreted before the work. The most extensive and conclusive experiments on this subject, however, are those performed upon the pedestrian Weston, by Prof. A. Flint, Jr.[1] In these observations the greatest care was taken in all the details. The amount of nitrogen in the food was calculated as well as that excreted, in order to avoid any error which might arise from the variation in the diet. Weston was under observation fifteen days—five days before, five days during and five days after the walk. The result was that the increase of urea discharged during the period of exertion was very marked.

These results have very recently received additional support by experiments upon the same individual, during a recent trial of endurance in London, conducted by Dr. Pavy,[2] who had never accepted Prof. Flint's conclusions, but who, on the contrary, held quite opposite views. It then appears settled that nitrogenous food and muscular activity increase the discharge of urea. But whether in the case of food the result is due to an excess of nitrogenous material in the blood being directly transformed into the excretion or whether the tissues are stimulated to more active metamorphosis by this excess, is a question as yet undecided.

The excretion of urea varies with the different periods of the day, being less in the morning and greatest at night. Indeed we have seen specimens of urine passed late in the afternoon and evening which

[1] New York Medical Journal, June, 1871, p. 669.
[2] London Lancet, 1876. Vol. I., No. ix. et seq.

gave an abundant precipitate with nitric acid (urea nitrate) without previous manipulation.

As a general rule an excessive discharge of water by the kidneys is accompanied by an augmentation in the amount of urea, and a diminished flow by a decrease.

Urea is formed and excreted as long as the vital functions are performed, whatever be the diet, and even when all food is withheld.

In *disease* the production of urea may be both increased and diminished. It is increased in most febrile and inflammatory diseases such as pneumonia, peritonitis, etc., but during convalescence sinks below the normal average. In diabetes mellitus, where the patient takes enormous quantities of food, the total daily quantity of urea may be 60 or 70 grammes, (900 or 1050 grs). In diabetes insipidus it is also increased, evidently on account of the excessive amount of water ingested.

The excretion of urea is diminished in renal diseases, cholera, phthisis and all other chronic affections accompanied by impaired nutrition. It is, however, evident that these deviations are not all due to the same cause. In diseases which interfere with the functions of the kidneys, the same amount of urea may be formed as in health, but the kidneys being impaired it accumulates in the circulation; while in diseases characterized by a chronic torpidity of the vital forces the metamorphosis of the tissues and food is retarded—but the small amount produced is properly drained away from the blood by the kidneys.

In Bright's disease the failure of the urea to be

separated from the blood causes it to collect there in great quantity, and also to appear in every tissue and fluid of the body. Thus confined it acts as a poison upon the nervous system inducing what is known as *uræmia*. Retained any length of time in the urinary organs urea is decomposed into ammonium carbonate, which is absorbed and gives rise to a diseased condition called *ammonæmia*. Whether urea is thus changed in the blood and uræmia and ammonæmia are identical is not known. The method of estimating the quantity of urea will be given in another place.

CREATININE,—$C_4H_7N_3O$.

This is another excretory product closely resembling urea in that it is crystallizable, contains nitrogen and has the properties of an organic base. Very little is known concerning it further than that it exists in small amount in muscle, and that it can be obtained from creatine of muscle, by the separation of the elements of water, as follows:

$$\underset{\text{Creatine.}}{C_4H_9N_3O_2} - \underset{\text{Water.}}{H_2O} = \underset{\text{Creatinine.}}{C_4H_7N_3O}.$$

Both creatine and creatinine exist in the urine to a very small extent. They are probably intermediate formations in the metamorphosis of nitrogenous matters towards urea, uric acid, carbonic acid and water.

The remaining ingredients of the urine which have a practical interest will be considered under "urinary deposits."

Color.—The tint of healthy urine is liable to variation, though a yellow amber is about the standard color.

The coloring matter of urine is a peculiar organic constituent which though diligently studied is not yet perfectly understood. Its proneness to decomposition renders it very difficult of separation in a pure state, and on this account a variety of substances have been described, differing somewhat in color and general characters, and christened accordingly. This is the reason we see so many names applied to the urine pigment.

The most important results were obtained by Harley,[1] and Thudichum.[2] Harley succeeded in obtaining a "bright red, non-crystallizable compound" which when fresh closely resembles red sealing wax. After a while it gradually becomes darker and more brittle, but does not lose its general characters even after twelve years keeping. He describes it as being insoluble in water, soluble in alcohol, chloroform and ether with a rich port wine color, according to the amount dissolved, and soluble in fresh urine imparting the characteristic tint of normal high colored specimens of that fluid. To this substance he gave the name *Urohæmatine;* for he believed it to be but the altered and effete blood pigment, and its proportion in the urine to be directly connected with the metamorphosis of that substance. Urohæmatine contains both nitrogen and iron. According to its discoverer it is this material which is set free and manifests itself in the darkening of color when nitric or any strong mineral acid is added to urine.

[1] The Urine and its Derangements. Phila., 1872, p. 96.
[2] British Med. Journal, Nov. 5, 1864.

ANALYSIS OF THE URINE. 15

Urochrome is the name by which Thudichum designates the pigment which he succeeded in extracting from the urine, and it is perhaps the best term to remember. When separated, it presents itself as a yellow, uncrystallizable mass; very soluble in water, to which it gives a yellow, urine-color. Whichever the pigment is, it is present in the urine in something like definite amount and depends upon the proportion of water whether it imparts a deep or faint color.

Excessive indulgence in water, malt liquors and wine, and diminished activity of the perspiratory apparatus, will cause an increase of the watery element in the urine, and consequently more or less dilution. Under these circumstances we would find the color light; in some cases resembling pure water. The converse of these conditions will produce a contrary result. Urine voided by nervous and hysterical patients is in large quantity and almost devoid of color.

Transparency.—Normal urine is perfectly clear, with the exception of a small collection of mucus and epithelium which nearly always collect at the bottom of the vessel, but may entangle a number of air-bubbles and float as a feathery ball just below the surface.

Reaction.—Healthy urine has an acid reaction, which is due not to the presence of any free acid, but to an acid salt, sodium bi-phosphate. This reaction may be decided or faint. But it should not be understood that unless the urine is acid it is abnormal. Indeed, the reaction may vary within healthy limits, between well-marked alkalinity and acidity; the irregularity being to a great extent due to diet.

The urine of carnivorous animals is acid, that of

herbivora is alkaline, while in omnivora the reaction can be said to occupy a position between the two. It has been ascertained that a vegetable diet will cause the disappearance of the acidity in the carnivora, and a regimen of flesh induce an acid condition in the urine of herbivora. Consequently in omnivorous man we are not surprised at the unstable reaction presented by his urine, and can account for the changes by reference to his food.

According to most observers the urine is alkaline after a mixed meal, but soon begins to grow less so, until the acid state is restored. The acidity increases during fasting, and reaches its greatest degree of intensity after about twelve hours abstinence from food. (Roberts.)

It is well to note that this variation is not due to any diminution in the acid salt present, but simply to a sudden accumulation of alkaline ingredients, the result of digestion. For during this process, all the salts of the organic acids, such as tartrates, malates, citrates, and lactates are transformed into alkaline carbonates, and as such appear in the urine. Substances containing these organic compounds are in daily use, either as food or medicine. Many fruits and vegetables offer familiar examples, and carbonated mineral waters are a common cause of a temporarily alkaline urine.

Specific Gravity.—The specific gravity of normal urine varies from 1018 to 1025. These limits may be extended in individual cases, and in fact, depend greatly upon the quantity of urine voided. In other words, the specific gravity bears an inverse ratio to the daily quantity. Should we have an increased flow of urine from unusual indulgence in drink, or from the action of any diuretic, we would not expect the density to equal that where the same quantity of solid material is present in a less amount of water. For in diuresis, except in a few instances of disease, we merely have the water increased, and not the solid constituents. In such cases we should not be surprised to find the specific gravity as low as 1010 or even 1005. On the other hand, where the individual has taken little or no fluid, or has perspired freely, or is suffering from diarrhœa, the urine would be what we call concentrated —that is, the normal amount of solid ingredients, salts, urea, etc., would be there dissolved in a small proportion of water. And under these circumstances, the specific gravity would be high—even 1030 would not indicate disease.

Daily Quantity.—The total quantity of urine voided by a healthy individual during twenty-four hours is estimated to be about 1200 cubic centimetres. This is subject to variation, depending upon the quantity of fluids drunk, the activity of the perspiratory functions, etc.; for it is evident that should there be a small proportion of fluids taken into the system, there will be less secreted by the kidneys, and *vice versa*. And should perspiration prevail to an unnatural degree, we would be getting rid of the water by another channel, and would not expect to find the same volume of urine; the same may be said of watery discharges from the bowels. The kidneys act as regulators of the water-supply of the blood; they take from it any excess, and when there is an insufficiency, they demand only enough to dissolve the solid constituents of the urine, and to facilitate their discharge from the body.

The estimation of the daily secretion of urine is one of the most important points connected with its study. But it will readily be seen, from what has been said, that it depends greatly upon the *specific gravity* whether the quantity voided has a clinical significance. In fact the daily quantity and specific gravity of urine are so closely related that it is difficult to treat of them separately.

The solids, urea especially, are increased under the influence of muscular exercise, and an animal diet; while the watery element is augmented by indulgence in drink, the action of diuretic medicines and by nervous or hysterical conditions. Variations in specific gravity, ranging from 1005 to 1030, cannot be *constant*, and not excite suspicion regarding the integrity of the kidneys, or some pathological condition of the economy. To entitle them to be considered under the head of normal, they must be only temporary and easily referred to some such cause as has been mentioned.

2. EFFECTS OF REAGENTS UPON NORMAL URINE.

Cold.—Cold has no visible effect upon urine of a specific gravity at or below about 1020. But in concentrated specimens, after cooling, there will be a precipitate, more or less colored, which consists of the amorphous urates, they being only soluble at an elevated temperature and in an excess of water. This precipitate first appears as a cloud throughout the whole volume of urine, but will gradually collect at the bottom and adhere in specks to the sides of the vessel as a fine powder.

Such a deposit or cloudiness will disappear upon again raising the temperature to that of the body.

Heat.—Normal urine of a decided acid reaction is unaffected by the application of heat. But should the reaction be faintly acid, neutral, or alkaline, heat will cause a cloudiness, due to the precipitation of the earthy phosphates of lime and magnesium.

These two phosphates are insoluble in a neutral or alkaline fluid, and are less soluble in warm than in a cold medium, and therefore will be precipitated by heat if the reaction of urine is even slightly acid.

Acid.—Vegetable acids have no immediate visible effect upon healthy urine; but strong mineral acids deepen the color. If nitric or hydrochloric acid be added, in the proportion of about one-fourth, after several hours minute but distinct dark brown crystals will be seen clinging to the sides of the test-tube or vessel, and collected at the bottom. These are the crystals of uric acid which have resulted from a decomposition of the urates by the acid. If the specimen be concentrated, voided after abstinence, a copious precipitate of urea nitrate in the form of white shining scales will follow the addition of nitric acid.

Alkalies.—When urine is rendered alkaline, the earthy phosphates of lime and magnesium will be precipitated.

Silver nitrate solution added to urine precipitates the chlorides. But it is necessary to previously acidify the urine with a few drops of nitric acid, otherwise silver phosphate will come down in addition to the chloride.

It is sometimes expedient to estimate the comparative quantity of the chlorides, inasmuch as it is pretty well established that they are subject to great fluctuations in certain forms of disease. In pneumonia, pleurisy, and cholera, the chlorides almost disappear from the urine, and the silver salt produces little or no effect. But when the attack begins to subside, this reagent will detect the return of the chlorides. The precipitate is a very copious white cloud which turns black on exposure to the light.

Barium chloride solution throws down the sulphates from the urine. It is well first to add a little nitric or hydrochloric acid to the urine as some barium salts, other than the sulphates, are thereby dissolved.

The variations in the daily excretion of the sulphates have little clinical significance as far as known.

Basic acetate of lead and **silver nitrate** precipitate the mucus and coloring matter.

Many substances taken by the mouth or injected into the circulation subsequently make their appearance in the urine. Such are *potassium ferrocyanide* which can be detected a few minutes after administration by the blue color which it strikes with ferric nitrate; *iodine* given either in the free form or in combination is easily shown to be excreted by the kidneys. It, however, never appears free in the urine, and therefore will not give the characteristic reaction with starch until liberated by a drop or two of nitric acid; *quinine* passes out by the urine and many articles impart their peculiar odor, as *ether*, *cubebs*, *copaiba*, *turpentine*, *peppermint*, etc.

CHANGES WHICH TAKE PLACE IN URINE AFTER BEING DISCHARGED FROM THE BODY.

If a specimen of urine be kept for observation, the following changes show themselves:

At first, after a period varying from two days to a week, the acidity becomes more marked, the color is darker, and crystals of *uric acid* and *lime oxalate* make their appearance. Even should the urine be faintly alkaline, to commence with, and cloudiness exist from the precipitation of the earthy phosphates, it will become acid, and the cloudiness clear up, when the process continues as in the other case. The acidity increases up to a certain point, and then begins to grow less. It may continue two, four, or seven days; and Lehman states, he has observed the acid reaction to increase for two or three weeks, and then not disappear altogether until eight weeks. This is the ACID FERMENTATION of urine (Fig. 1), which although not constant, will be observed in the majority of cases.

Fig. 1.

Acid Fermentation: Uric acid, octahedra of lime oxalate, and amorphous urates.

The acidity gradually becomes fainter and fainter, and at length the urine is neutral. Now, marked changes will be noticeable. The urine is cloudy; myriads of vegetable organisms in rapid motion are visible under the microscope (Bacteria). Neutrality gives place very soon to alkalinity, which is advertised most emphatically by a putrescent, ammoniacal odor. The uric acid crystals disappear, and others of the *ammonio-magnesium phosphate* are produced. These crystals are large prisms, and can be seen glistening on the surface of the urine where a layer of brittle fatty matter has formed. Dark, round crystals of *ammonium urate* may also be seen with the microscope.

This state of things continues until the decomposition is complete. . Then the ammoniacal odor will no longer be detected, and the urine will have lost most of its color.

This constitutes the ALKALINE FERMENTATION which is constant (Fig. 2)

The *chemistry* of these two fermentations has been differently explained; but that offered by Scherer is generally accepted as correct.

He says that the organic matters, the coloring matter and mucus, act as catalytic bodies, and induce the fermentation whereby lactic acid is produced. Just how lactic is formed is not well understood, but it is certain that it makes its appearance in urine when allowed to stand,

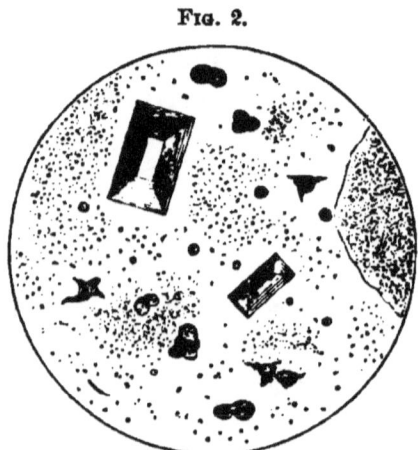

Fig. 2.

Alkaline Fermentation: Crystals of ammonio-magnesium phosphate, ammonium urate and deposit of amorphous lime phosphate.

in which it could not be detected when fresh. The presence of lactic acid then decomposes the urates, whereby uric acid is set free and makes its appearance as crystals. At the same time, oxalic acid must be produced, for crystals of the oxalate of lime show themselves. This oxalate of lime is very insoluble, and could not have existed in the urine before without detection. Therefore the supposition is that oxalic acid is formed and immediately unites with the lime already present, for which it has a great affinity. Theoretically we can account for the production of oxalic acid from uric acid. For when this is subject to oxidizing agents it is decomposed, among other substances, into oxalic acid. So much for the acid fermentation.

Now, we have seen (p. 6) that urea is converted into ammonium carbonate when in contact with a decomposing organic substance, by the addition of two equivalents of water.

Here then in the urine we have the conditions of this decomposition. The mucus gradually loses its power of producing lactic and oxalic acids, and begins itself to decompose; and surrounded by water, the arrangement for the transformation is complete. It accordingly takes place, and the first effect of the presence of ammonium carbonate is to neutralize the acid reaction, and then to induce the alkaline. Now, of course, the earthy phosphates are no longer soluble, and render the urine opalescent by their precipitation; the uric acid crystals are dissolved; a scum of animal matter intermixed with the amorphous phosphates forms on the surface, and very soon glistens with crystals of a new formation, *ammonio-magnesium phosphate.*

The urea continues to be decomposed, and ammonium carbonate to unite and form these several new substances, until there is nothing more for it to combine with. Now it escapes as gas, and the odor reminds us of that common to public urinals, where, indeed, the same process as just described is going on. At length, all the urea is decomposed, and the evolution of ammonia ceases. The other substances either remain unchanged or pass off in the form of other gases.

These are the important facts concerning normal urine, and a knowledge of them is indispensable, in that if we are not acquainted with them, we certainly shall not be prepared to detect and appreciate the variations which constantly present themselves, and constitute an abnormal condition.

PART II.

CHARACTERS OF ABNORMAL URINE.

Odor.—The odor of urine is frequently affected, and is likely to attract the attention of both patient and physician, and lead to its examination. Many articles taken as food and medicine impart to it an odor peculiar to themselves. Such are asparagus, onions, turpentine, cubebs, and copaiba. An ammoniacal odor tells the story of decomposition, and urine containing pus, blood or albumen very soon decomposes and emits a putrid odor.

Color.—The color of urine is subject to many changes depending simply upon the degree of concentration. In febrile diseases, where we have a partial suppression of the watery element, the secretion is high-colored. In diabetes mellitus, a disease characterized by an inordinate flow of urine of high specific gravity, and containing sugar, the tint is light and of a peculiar straw color. In diabetes insipidus the urine sometimes resembles pure water. In albuminuria, especially of long standing, there is a peculiar whitish, albuminous appearance which is highly characteristic. But we should never rely upon the color as indicating the presence or absence of albumen.

Certain articles of food and medicine affect the color of the urine. Strong coffee heightens, rhubarb imparts a deep yellow, and logwood gives it a reddish hue. Santonine renders it an orange red when alkaline; when acid, a golden yellow. Creosote and compounds of tar have been known to cause the urine to become almost black; and, lastly, blood and bile may be present in such quantities as to be readily recognized.

The presence of bile may be distinguished from the effects of rhubarb by the addition of a little liquor ammonia, when the deep yellow of the latter will be converted into a crimson.

Blue, green, and *black* urine is occasionally seen. These remarkable pigments occurring in this situation were long in being understood. But it is now generally acceded that they are due to the presence of *indican*, a colorless substance, identical with the vegetable product from which the indigo-blue and indigo-red of the arts are obtained; and that they represent but different stages of oxidation of this substance. These peculiar colors never appear in freshly voided urine, but are always observed in that which has been allowed to stand exposed to the air. It depends upon the amount of the coloring-matter whether the entire volume of the fluid will be affected. Sometimes its presence is only manifested by the affinity which it has for small solid particles, such as epithelium debris, crystals and extraneous matters, when these substances will be seen under the microscope deeply stained red, blue, etc.

The addition of mineral acids will precipitate these pigments in urine in which indican exists.

Whether these extraordinary appearances have any

clinical significance is not decided. Indican has been detected in many specimens of healthy urines.[1]

Transparency.—If a specimen of urine under examination is not clear and transparent, it should be first ascertained whether there was any turbidity when first voided; for we have seen that perfectly healthy urine, when kept for any length of time, will undergo changes and a marked opalescence exist. And abnormal urine may be at first perfectly clear, but on cooling or standing exhibit a cloudiness or deposit. It is possible for urine to be abnormal, and yet remain free from turbidity or deposit. For example, albumen and sugar are perfectly soluble, and can not be detected by simple ocular inspection.

Substances which interfere with the transparency of recently discharged urine are, *pus*, *blood*, *phosphates*, *urates* (if the specimen is cool), *chyle*, *spermatozoa*, and *epithelium*.

The reader is referred to each of these under the head of "Urinary Deposits."

Reaction.—After what has been said concerning the variability in the reaction of normal urine, it is sufficient to add now that when we meet with a patient whose urine is habitually or most of the time neutral or alkaline, we should regard him as a subject for treatment.

Remember that the alkalinity may be due to the presence of the fixed salts of sodium and potassium, or

---

[1] Literature: Hassall, Philosophical Transactions—1864, p. 297. Harley. "Urine and its Derangements." Phila., 1872, p. 107. In Thudichum, "On the Pathology of the Urine," London, 1853, p. 328, a very full account is given.

to the volatile one of ammonium; that the former are derived from the blood and the latter is the result of the decomposition of urea. These two conditions can easily be distinguished apart by the ammoniacal odor which betrays the presence of ammonium carbonate, and the fact that the blue color which it imparts to reddened litmus-paper fades away. And also, if the reaction be due to this alkali, we shall find crystals of the ammonio-magnesian phosphate under the microscope. (Fig. 6.) Where either the fixed salts or the volatile one is present, of course the urine will be turbid from a precipitation of the earthy phosphates. And this precipitation occurring in the bladder is likely to give rise to a calculus; yet it is pretty well established that the urine can remain alkaline a long time from sodium and potassium carbonates, and a stone not form. It is crystalline deposits which we have to fear in this regard, and as the volatile alkali induces crystalline formations, calculi or concretions are apt to accompany it.

Then, too, we may have the urine abnormally acid, producing a scalding sensation during micturition. Now, we know what the result is when an acid is added to the urine outside the body: uric acid crystals will be precipitated. So it is in the bladder under the same conditions. These crystals then accumulate and form gravel, which may lodge in the kidney, ureter or bladder. In either situation it is easy to comprehend how their presence and increase in size will cause retention and decomposition of the urine and growth of the original uric acid concretion by successive layers of earthy phosphates.

Daily Quantity and Specific Gravity.—We have seen, in the case of healthy urine, how closely the specific gravity and daily quantity are related, and that as one increased, the other decreased; or, in other words, that they bore an inverse ratio to each other. In certain diseased conditions, however, both are increased and decreased together. The specific gravity may be 1030, and the daily quantity 2 litres (60 oz). Here we would recognize the fact that there is a double waste of both water and solids. And, again, the specific gravity may be 1006 to 1012, and the quantity not exceed 400 cubic centimetres (12 oz). In this case, there is a suppression of both elements, and from retention of the chief solid constituent, urea, the most fatal results may follow. There is also a disease —diabetes insipidus—where the daily quantity of urine is enormous, and the specific gravity much *below* the normal standard. This state of things robs the system of its proportion of water, occasioning great thirst, whereby nature endeavors to counteract the drain.

Here, then, as a general rule in abnormal urine, the specific gravity and the daily quantity are in direct ratio. *When the amount of water is increased, the solids will be also, and vice versa.*

The specific gravity of urine will be increased by abnormal ingredients, such as pus, blood, mucus, albumen, and sugar.

Great care is necessary in collecting urine for examination. The patient must be made to pass the entire secretion of twenty-four hours in one clean vessel, and from this total quantity a portion must be selected for examination, especially as regards the specific gravity.

PART III.

1. URINARY DEPOSITS.

THE urine is subject to deposits, or collections of solid and semi-solid substances, which, on account of their weight, subside when undisturbed. These deposits consist of various materials, some of which are normal constituents of the urine, either separately or in combination, while others are foreign to its composition altogether.

It will be convenient to study urinary deposits by dividing them into these two classes.

TO THE FIRST CLASS BELONG URIC ACID, THE URATES, HIPPURIC ACID, THE PHOSPHATES, LIME OXALATE, EPITHELIUM, MUCUS, AND PIGMENTS.

Uric Acid.—$C_5H_4N_4O_3$. This is another nitrogenous excrementitious substance very closely resembling urea; its chief difference being that it is not so prone to decomposition, and does not exist in the body in a free state, under normal conditions.

Uric acid as soon as formed unites with a portion of the alkaline bases of the phosphates in the blood, and appears in the urine as sodium and potassium urates.

It appropriates the sodium base in greater proportion and thus results the formation of *sodium biphosphate* and the acid reaction of the urine:

Sodium phosphate.   *Uric acid.*   *Sodium urate.*   *Sodium biphosphate.*

$$Na_2HPO_4 + C_5H_4N_4O_3 = NaC_5H_3N_4O_3 + NaH_2PO_4.$$

By oxidation uric acid is converted into urea and oxalic acid.

There are various pathological conditions, not well understood, which give rise to free uric acid in the urine. Slight derangements in the digestive process, an excess of nitrogenous food or a want of alkalies will cause it to appear. When free it invariably presents itself as crystals, in most instances of a brown color. We have seen that uric acid is set free during the acid fermentation of urine, and that it is caused by the development and action of lactic acid, which decomposes the urates. We can accomplish the same result by adding any strong acid to a test-tube containing urine, and allowing it to remain quiet for several hours. Then the characteristic uric-acid crystals will be seen attached to the sides and collected at the bottom.

The dark brown color which almost invariably distinguishes these crystals is derived from the coloring matter of the urine, for which they have a great affinity. Yet we sometimes see almost colorless and pure crystals of uric acid; and these are apt to be small square plates and diamonds.

The formation of crystals of uric acid may take place in the urinary passages, not only as a result of fermentation, but spontaneously, as just indicated. In either case, gravel and calculi

Fig. 3.

Uric Acid.

are likely to result. It is very important to know whether a deposit in a specimen under observation has taken place previous or subsequent to its discharge.

Fig. 4.
Uric Acid.

A deposit of uric acid is of a dark brown color, or if pure, it is a white, glistening sediment, and under the microscope presents the greatest variety of crystalline forms. In fact, so numerous are the shapes and arrangements of these crystals, that it is very difficult to describe them. Yet when familiar with urinary crystals, they are not likely to confuse one. For it is only necessary to be able to recognize the other crystalline deposits, which vary little or none; and then when a brown crystal of an unusual form is seen, it is pretty safe to pronounce it uric acid.[1] The most common microscopic appearances are represented in Figs. 3 and 4.

TESTS.—Uric acid is insoluble in water, alcohol, and ether. It is soluble in an alkali, especially at a high temperature, and in sulphuric acid without decomposition, for it can be reprecipitated simply by the addition of water.

There is a beautiful test for uric acid called the

[1] Where any doubt exists as to the identity of a colored crystal it should be dissolved in a little warm alkali and then treated with hydrochloric acid; when if it be uric acid minute colorless plates will be detected after some hours, under the microscope.

murexid test. Place the uric acid in a clean white porcelain capsule, and add a drop or two of nitric acid; then evaporate over the flame of a spirit-lamp. When the nitric acid has been driven off, there will a pink stain show itself, which on the addition of a little liquor ammonia, assumes a beautiful purple color.

Urates.—We have just seen that uric acid, simultaneously with its formation, unites with a part of the loosely combined sodium and potassium of the phosphates, and forms sodium and potassium urates. (It is a question whether the urine contains lime and ammonium urates.) The urates constitute the most frequent deposit met with in the urine. They appear as a dense cloud, which collects at the bottom of the urine-glass, and at the same time dusts the sides over with a powdery film. The color of the precipitate varies from a white to a red, according to the concentration of the urine, these substances having a great affinity for the coloring matter.

The urates have very much the same characters, being soluble in an excess of water at a low temperature, and soluble in very small proportion of this medium under the influence of heat. It is this reason that, however concentrated the urine may be when voided, it is never turbid with the urates; but in a short time, the temperature falling, a cloudiness is visible, followed shortly by a copious deposit.

Knowing, then, that the urates are precipitated when the urine is deficient in water, we can understand how abstinence from drink, profuse perspiration, a watery discharge from the bowels, or any influence which lessens the normal quantity of water in the system, or turns it away from the kidneys, will have the effect of concentrating the urinary secretion, and causing the urates to appear when the urine cools.

Yet, a deposit of the urates is not on all occasions due to a suppression of the watery element; there are various disturbances of the system wherein their production and excretion is enhanced.

In fevers, heart and liver diseases, the urates are deposited and are high colored. Irregularity in the digestive apparatus often induces their precipitation, and we often encounter a copious white deposit of them in the urine of teething children.

A deposit of urates is almost always amorphous, though sometimes crystals of sodium urate are formed, and in the alkaline fermentation, ammonium urate. Crystals of sodium urate are met with in the chalky concretion common in the bodies of gouty persons, and also in the urine of such. In the concretions they are needle-shaped, arranged in rosettes of various degrees of perfection: while in the urine they assume a globular form, with irregular projections along the surface. (Fig. 5, *b*.)

Ammonium urate crystals appear as very dark, spherical masses, and as dumb-bells. They form during the alkaline decomposition, and can only be seen under that condition. (Fig. 5, *a*.) The

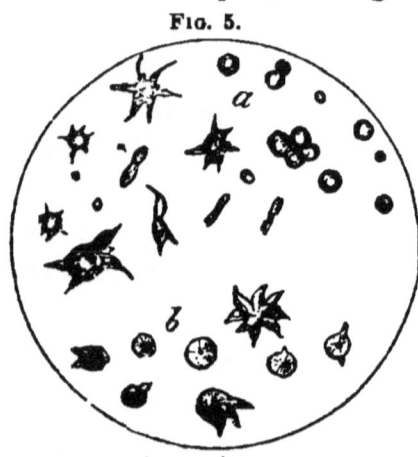

Fig. 5.
a. Ammonium urate.
b. Sodium urate.

pathological indications of the crystalline urates are much more important than those attaching to the amorphous deposit, in that the former are deposited in the urinary tract, while the latter seldom are.

A urine showing a precipitation of urates is always acid.

TESTS.—The urates disappear at a temperature of 100 F. They are soluble in an alkali. The addition of an acid decomposes them, with the liberation of uric acid which very soon crystallizes. This process is very interesting to observe under the microscope.

The murexid test for uric acid is applicable to the urates.

Hippuric Acid.—$C_9H_9NO_3$. Hippuric acid was formerly believed to exist only in the urine of herbivora; but it is now generally conceded that it is quite constant, in small proportion, in the renal excretion of man. It increases under the influence of a vegetable diet, and can be made to appear in large amount by taking benzoic acid by the mouth—this substance being transformed into hippuric acid. Hippuric has very similar characters to uric acid, and their compounds possess no important distinctions.

The hippurates owing to their solubility never appear as deposits in urine. But they are readily decomposed by acids, when the *hippuric acid* will form a crystalline deposit in very concentrated solutions. Crystals of hippuric acid may then be encountered during the acid fermentation of urine or as a result of the addition of an acid in our manipulations. The acid is easily obtained from the urine of the horse, or by administering a quantity of benzoic acid to a man or dog. Then upon evaporating and adding hydrochloric acid, a visible deposit of hippuric acid is obtained, which under the microscope is seen to consist of long transparent prisms and needles.

TESTS.—These crystals may be mistaken for those of the phosphates. In such a case a drop of hydrochloric acid will decide the question; the phosphates being thereby dissolved and hippuric acid unaffected.

To the taste they are bitter. The watery solution reddens litmus paper, showing it to be a stronger acid than uric.

Phosphates.—The alkaline phosphates of sodium and potassium, owing to their solubility, are never met with as deposits in urine.

It is the earthy phosphates of lime and magnesium with which we have to deal as precipitates, and which are so frequently encountered. They present themselves in three forms: *the amorphous lime phosphate, the crystalline lime phosphate, and the ammonio-magnesium phosphate.*

Amorphous lime phosphate, with a small amount of magnesium phosphate, is the common deposit of urine rendered alkaline by a vegetable diet or the administration of vegetable acids and their salts. It frequently becomes abundant after nervous exhaustion, as loss of sleep, mental application, etc.

This deposit renders the urine milky when voided, and subsequently subsides as a not very abundant white collection at the bottom of the vessel.

Crystalline lime phosphate is sometimes met with.

The crystals are perfectly colorless needles arranged in radiating bundles. They are sometimes called the *stellar phosphate.*

Roberts considers their presence indicative of some grave disorder, having discovered them in diabetes, cancer, and phthisis. It may, however, appear in normal urine when much lime has been introduced into the system.

The ammonio-magnesian phosphate is a crystalline deposit, the result almost always of the alkaline fermentation and the decomposition of urea. This deposit very seldom takes place inside the body, and when it does, it is the result of retention of urine. Under these circumstances, decomposition takes place, and the ammonium, which is one of the results, unites with the magnesium already present, and the deposit is formed. The ammonium normally present in the urine is seldom sufficient to constitute this deposit *without the decomposition of urea.*

The earthy phosphates are only soluble in an acid fluid, the acid salt sodium biphosphate holding them in solution in the urine. Whenever the urine becomes neutral or alkaline from a diminution in the quantity of uric acid or from an excess of alkaline ingredients in the blood, or from retention and consequent decomposition, the earthy phosphates will be thrown down.

Notwithstanding the great amount of literature on the subject, "phosphaturia," with the exception of that induced by the alkaline decomposition, has not much clinical significance. Under the microscope the phosphates appear as amorphous granules and, in the case of the ammonio-magnesian, large glistening prismatic crystals. When crystallization takes place rapidly we often see beautiful feathery forms like those represented in the lower portion of the accompanying figure.

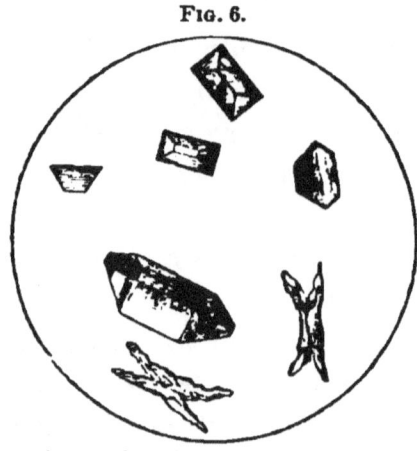
Fig. 6.
Ammonio-magnesian phosphate.

Tests.— The earthy phosphates are soluble in an acid solution; and are less soluble in hot than cold media. A drop or two of any acid will cause them to disappear.

Lime Oxalate.—Lime oxalate is found in urine as a result of the acid fermentation, and also in perfectly fresh and undecomposed specimens. Indeed, the formation may take place in the body. It is claimed by many that oxalic acid is a normal constituent of the urine, and that it, like the other ingredients, is subject to variation, and when furnished in a little more than the usual amount, it unites with the lime already present, and the oxalate is the result. It is probably derived from uric acid by oxidation.

It appears as a whitish powder, and never is very abundant.

A great deal has been said concerning the significance of oxalate of lime in the urine, under the heads of "Oxaluria" and "Oxalic Diathesis."

A long train of symptoms were attributed to its presence in the urine, such as nervous prostration, despondency, loss of sexual power in the male, etc.

And although there is not as much interest in the subject now as formerly, I have reason to believe that the doctrine is not altogether false.

But the chief significance which attaches to this deposit is the possibility of the formation of a stone; and as this variety is very hard and rough, it is especially dreaded.

Tests.—A deposit of lime oxalate is always crystalline, and the microscopic appearances are sufficient to distinguish it. The crystals are of two kinds, octahedra and dumb-bells, each being colorless and transparent.

The octahedra, which are by far the most common, consist of two pyramids placed base to base and varying in size from about $\frac{1}{1000}$ of an inch to that which may be seen by the unaided eye.

The dumb-bells are highly characteristic. Occasionally an oval mass is seen which appears to be a dumb-bell in process of formation.

Lime oxalate is soluble in strong mineral acids.

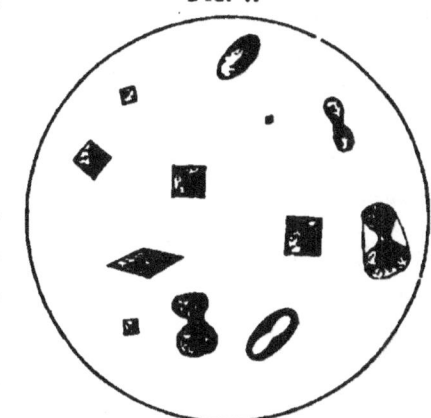

Fig. 7.

Lime oxalate: Octahedra and dumb-bells.

Epithelium.—Every specimen of urine contains some epithelium, which in certain diseases of the kidneys and urinary passages is greatly increased. In the healthy state, we find that from the bladder in the greatest amount; and in the female, vaginal epithelium will be mingled with the urine, if great care is not exercised.

As the genito-urinary apparatus is lined with epithelium of different forms throughout its extent, it is well to know what are the peculiarities which distinguish that of one region from that of another.

Vaginal epithelium is very common in the urine of women, especially when there is any discharge from the vagina or uterus. It is easily recognized by its large size, thinness, wavy outline, and disposition to fold upon itself. (Fig. 8, *b*.) The female urethra contains very similar forms, but they are somewhat smaller.

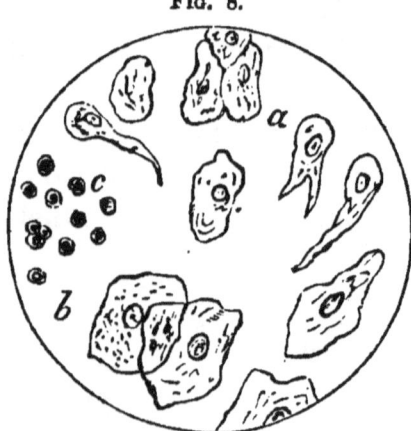

Fig. 8.

Epithelium: *a*, Bladder; *b*, Vaginal; *c*, Renal.

The urethra is lined by epithelium, which differs in the spongy and prostatic portions. From the meatus to the prostatic region, the cells are round and oval. About the prostate, they are spindle-shaped, caudate, and irregular.

In the *bladder*, the size is increased, and they are often seen still united by their edges. (Fig. 8, *a*.) Those from the ureters possess the same characters, only are of smaller size.

The pelvis of the kidney contains small round and oval cells, of the flat variety, which are a little larger than tubular epithelium.

The epithelium of the kidney tubules is a very

complicated one, but it is sufficient for our purposes to know that it is spherical, about twice as large as a blood-globule, from which it may be distinguished, should any doubt arise, by its nucleus. It usually comes away from the kidney adhering to fibrinous moulds of the tubes, but may often be seen floating free in the urine. (Fig. 8, c.)

When epithelium is present in abundance, it will be accompanied by an increase in the natural amount of mucus, and the two substances settle to the bottom of the urine-glass, intimately mingled. Under the microscope, patches of epithelium-cells will be seen held together by the adhesive mucus.

Epithelium appears in the urine as a result of any inflammation or mechanical irritation of the mucous membrane of the urinary tract.

Mucus.—There is always a small amount of mucus in healthy urine, especially in the first passed in the morning. It is to the presence of mucus, however limited in quantity, that the decomposition of urine is due. For if the urine be filtered, it may be kept for an indefinite time without any change manifesting itself. (Scherer.)

Mucus appears as a deposit at the bottom of the glass, or may entangle air-bubbles and float just below the surface. It has the color of the urine containing it, and when separated on a filter, is perfectly transparent and glairy. All amorphous and crystalline deposits, blood, pus, casts, etc., become mingled with it and interfere with the transparency, or may mask its presence.

Excessive secretion of mucus may be the result of

mechanical irritation, ammoniacal decomposition of the urine in the bladder, or chronic inflammation. Sometimes there is such an amount produced that the whole volume of urine will be rendered semi-solid, and will rope like the white of egg.

Tests.—Mucus is liable to be confounded with pus, especially when colored by amorphous deposits. It may be distinguished from pus by its ropy and viscid nature, and, best of all, by its appearance under the microscope, mucus having no corpuscular elements.

Pigments.— We frequently encounter peculiar little bodies under the microscope, possessing indefinite shapes and appearances, and attracting the attention of the observer by their high color, being either red, dark brown, or yellow.

They sometimes resemble epithelium, but show no nucleus; again, they may be an irregular mass, unlike any thing we are familiar with. What their origin and significance is, is an unsettled question. They are likely to occur in any specimen of urine, but Roberts states that in several cases of chronic Bright's disease, he has noticed a great increase of them.

It has been thought that they might be epithelium-scales, stained by the coloring matter of the urine, as this is known to undergo changes whereby other colors are produced. The oxidation of indican probably has something to do with the matter.

One other source, and a common one, of these colored specks under the microscope, it is important to guard against: it is that they may exist in the glass slides or covers; for these articles are polished with a red powder, particles of which become imbedded in its minute irregularities.

II.

SECOND CLASS OF DEPOSITS: those which are foreign to the composition of the urine under any form.

This class differs from the preceding in that most of the deposits have a pathological significance.

The following are included under this head: BLOOD, CASTS, PUS, OIL, CHYLE, SPERMATOZOA, CYSTINE, KEISTINE, CONFERVOID VEGETATIONS, and BACTERIA.

Blood.—Blood may be mingled with the urine from either of the organs through which it has to pass. Urine containing blood may or may not give evidence of it when first passed. It depends upon the quantity of blood present whether there will be any distinction as regards color.

If the quantity is small, we probably will not suspect its presence until the urine has remained undisturbed some hours in a urine-glass, when a red line or layer will be discernible at the bottom. (It must not be confounded with uric acid, which is of a dark brown color, instead of a blood-red.) The deposit consists of the red corpuscles.

We may have urine stained with blood, and yet containing no corpuscles, or at least a very few. Thus the blood may be represented in the urine under these two conditions: *hæmaturia*, where the elements of the blood are present under their natural forms; *hæmatinuria*, where the red corpuscles appear to have disintegrated or dissolved, and only a few perfect ones can be found. It is necessary to consider these apart.

Hæmaturia.—In hæmaturia the red corpuscles are present, and impart to the urine a more or less red or smoky hue. The corpuscles have escaped from the circulation as a result of ruptured vessels somewhere in the urinary tract, which may be due to violence, as falls or blows, Bright's disease, temporary congestion, ulcers, abscess, and the lacerations attending the presence of stone and gravel. Hæmaturia is common in some irruptive fevers, notably scarlet; it is very frequently induced by the administration of turpentine and cantharides, and it may occur vicariously for menstruation and a hemorrhoidal flux. The urine of women is extremely liable to contain blood during the menstrual period, from admixture outside the body.

Hæmatinuria.—This curious affection is characterized by a chocolate-colored urine, due to the presence of hæmoglobin, which had transuded through the blood-vessels, and appears independent of the blood corpuscles which are thought to have dissolved. Dark granular and hyaline casts are frequently found, as well as amorphous matter and octahedra of lime oxalate.

The occurrence of the blood coloring matter in the urine independent of the corpuscles characterizes a disease which has recently received considerable attention. The chief features of hæmatinuria are first of all, its intermittent character; and second, the total absence, in the majority of cases, of symptoms attributable to disorders of the urinary apparatus.

ANALYSIS OF THE URINE. 45

In short, the general opinion is that it is a form of malarial poisoning, and it is consequently frequently spoken of as malarial or paroxysmal hæmaturia.

Whether there is a marked absence of the red corpuscles in this disease is a question upon which authors disagree. In a case under my own observation there was never any scarcity of them during the attacks. Dr. Legg[1] attributes the failure of some to find them to their rapid disappearance in urine when kept for some hours. The urine contains albumen during the paroxysms, but is generally entirely free from it in the intervals.

TESTS.—A rupture permitting the escape of the blood corpuscles will, of course, allow of the passage of albumen, and consequently we always find this substance in hæmaturia. The corpuscles themselves are albuminoid, and were there no free albumen present, would respond to the heat and nitric acid tests. But the most positive proof of the presence of blood is furnished by the appearance of the sediment under the microscope. This instrument, with a $\frac{1}{4}$-inch lens, will reveal the corpuscles. They are bi-concave discs, about $\frac{1}{3000}$ of an inch in diameter. By careful focusing, their form can be distinguished, the outline and centre never being equally distinct at the same time; then if a current be excited in the drop under the eye, by touching the side of the thin glass cover with a bit

---

[1] On Paroxysmal Hæmaturia: St. Bartholomew's Hospital Reports. London. 1874. Here a full *resumé* of the subject is given.

of blotting-paper, the corpuscles will roll over and over and over and show their thin edges and concave surfaces. All these points are illustrated in Fig. 9, a.

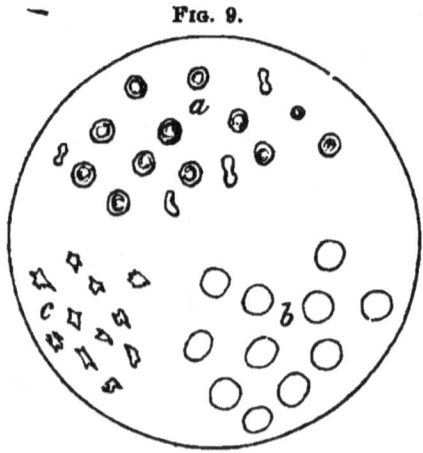

Blood Corpuscles: a, Normal; b, Swollen by absorption; c, Shriveled from lack of water.

There are changes, however, which blood corpuscles undergo when subjected to unnatural influences. In the blood serum, they maintain their shape and size, but immersed in a watery solution like urine, they immediately begin to absorb water, and very soon have swollen to twice their natural size, lost their color and bi-concave character, and have become spherical. When allowed to dry, they shrivel and become so irregular that should one not be familiar with the fact, he might not suspect their presence. When in this condition, the addition of a drop of water containing a little common salt will cause them to assume something of their original shape.

In the event that we should fail to detect any corpuscles when blood is suspected, they having dissolved or never having been present, it is only necessary to apply heat and nitric acid. Then if there be hæmoglobin present, it will assume a very dark color, become entangled with the coagulated albumen, and if allowed to rest, will collect at the bottom of the test-tube as a chocolate-colored mass, with the transparent urine above.

The spectroscope will detect exceedingly small quantities of hæmoglobin in solution, and should any doubt exist as to whether blood is present in a specimen of urine, this instrument will very readily determine.[1]

Casts.—Casts are moulds of the uriniferous tubules. The kidney is largely supplied with capillaries, which form a complex network about the tubules, and any congestion of the organ is very apt to result in an exudation from the blood-vessels into these little canals, of a peculiar substance, the composition of which is not well understood. This substance has the property of spontaneous coagulation, and thus adapts itself to the shape and size of the tube, and the urine collecting behind, washes it out, and it subsequently appears in the urine as a cast. The pressure which induces this exudation is also sufficient to cause a transudation of the liquid portion of the blood; and it results that *albuminuria* always exists where casts are formed.

Casts are not all of the same diameter. One reason for this is, that the tubules themselves vary in size in different parts of their course. Another cause depends upon the fact that in certain affections of the kidney, the epithelium lining the tubules is detached and voided with the urine. Now, a tubule thus denuded is larger than it was previous to the shedding of its lining, and consequently will afterward form a larger cast.

Casts differ as regards appearance. The coagulable

---

[1] See Dalton's Physiology. Phila.. 1875. P. 248. And a paper by the author, The Spectroscope; its Value in Medical Science. Trans. N. Y. Academy of Medicine. 1876.

matter which transudes into the tubules is perfectly transparent and structureless, so far as has been ascertained. The theory is, that this material when collected in the tubules fills and distends them; this distention causes pressure upon the epithelium lining which adheres to and sinks into it to a certain extent. Now, when the cast is washed out by the pressure of urine from behind, it pulls slightly upon its epithelial attachments, and if the cells be detached or detachable, through degeneration, they must come along with it.

It follows, then from the above description that if the kidney is not diseased to such an extent as to allow of desquamation of its epithelium, or, on the other hand, if the epithelium has been previously shed, leaving the tubule bare, the cast will come away and appear in the urine unaltered as regards its structure, and will appear as a transparent mould of a renal tubule. These we call *hyaline casts*.

It is not necessary, after the above explanation, to say more than that when the epithelium does show itself adhering to the cast, we recognize another variety—namely, *epithelium casts*.

And suppose these epithelium cells to have undergone a fatty degeneration, we should then have *fatty casts*. Go a little further, and imagine the epithelium to have suffered what is known as a granular degeneration, then we should find *granular casts*.

Besides these varieties, we meet with several others, having nothing to do, however, with the epithelium element. These are *blood*, *pus*, and *waxy casts*.

As it is my purpose to give a few hints respecting the practical bearing of the subjects treated of in this manual, it will be necessary to glance at each one of these casts separately.

Hyaline Casts, as their name implies, are structureless and transparent. They vary in diameter from the width of one blood corpuscle to that of three. Their length is not definite. Being transparent, they are difficult to find, and on this account are often overlooked.

The mode of formation of these casts has been stated above; and therefore when they are small, it is fair to suppose the kidney yet to be comparatively sound, and the cause of the symptoms which led to the examination of the urine, not of long standing. At any rate, we meet with small hyaline casts in acute Bright's disease. But what does a large hyaline cast signify? It tells the story of a more advanced disease, and its transparency is not due to a refusal of the epithelium to come off with it, and thereby asserting its healthiness; but, on the contrary, its very size bears evidence of the previous shedding of the epithelium, and of the present nudity of the tubule whence it came.

A cast of this last variety may be perfectly transparent, or it may be dotted over with a few specks of granular matter, and perhaps here and there a broken-down epithelium cell.

FIG. 11.

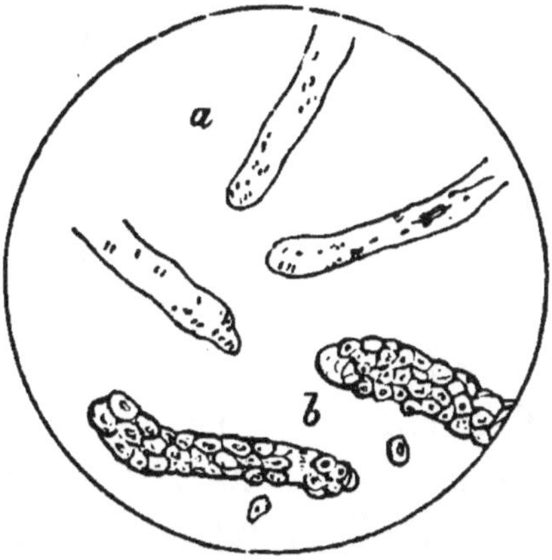

a, Hyaline casts; *b*, Epithelium casts.

It is evident that in chronic disease of the kidney, both small and large hyaline casts will be found in the urine, because the organ is not equally diseased in all its parts. (Fig. 11, a.)

Epithelium Casts.—We have seen that the effusion of fibrinous material into the uriniferous tubules undergoes spontaneous coagulation, and is afterward washed out by the urine which collects behind it; and that if the epithelium lining the tubules be detached or detachable, it will adhere to the mould, and be afterward found in that situation in the urine.

The cells may be scattered upon the surface of the cast or may present a regular arrangement, the same which existed while they yet occupied their normal place in the kidney. At the same time, there will be found free epithelium in the urine.

Epithelium casts are found in the urine of persons convalescing from scarlet fever, and in acute Bright's disease. In pneumonia and severe inflammatory diseases, they are the prevailing variety of cast. (Fig. 11, b.)

Fatty Casts are epithelium casts in which the epithelium has undergone a fatty degeneration. The cells appear to have been filled with fat and to have burst, discharging their oily contents into the tubules, where they have adhered to the fibrinous effusion. The cast will be more or less covered with these minute oil-drops and epithelial cells, which are also full of oil or fat. At the same time, fatty cells and free fat will be seen floating in the urine. (Fig. 12, a.)

Casts of this description are found in chronic affections of the kidney, and their presence in the urine is a very grave sign.

Granular Casts represent the epithelial element in a state of granular degeneration. What this granular material is, is not very well understood, though supposed to be fat in a state of fine subdivision. The cell may have entirely disappeared, or may remain, filled with these granules. These casts, appearing in abundance, are indicative of serious changes in the kidneys, and their presence determines an unfavorable prognosis. Fig. 12, *b*, illustrates the granular cast. It will be noticed that they are dark, and have ragged extremities.

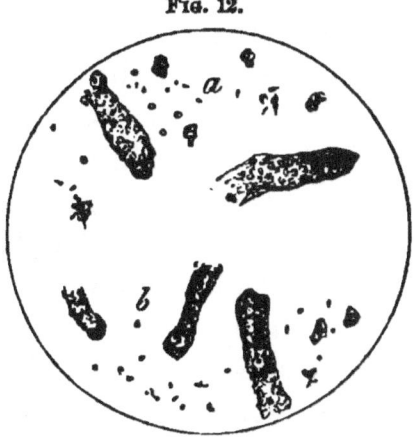

Fig. 12.

a, Fatty casts; *b*, Granular casts.

Blood Casts. — If there is a hemorrhage in the kidney when the conditions are present which induce the formation of casts, the blood-globules will attach themselves to the mould, sometimes very regularly, and form what is called a blood cast. At the same time, they will be present in a free state in the urine.

Blood casts are common in scarlet fever and any form of acute Bright's disease. They do not generally present themselves for any length of time, being the result of a hemorrhage which is not apt to continue.

Pus Casts.—These are rare. They are met with where the kidney is the seat of abscess. Yet cases where abscess was found at post-mortem examinations, and furnished no pus casts during life, are recorded. When present, they have the appearance of pus corpuscles adhering to the fibrinous cast. The urine will at the same time contain more or less pure pus. To prevent confounding pus corpuscles and epithelium, it is well to note that they are smaller, and the addition of acetic acid will render the cell-wall of the pus corpuscle transparent, and reveal a distinct granular nucleus.

Amyloid or Waxy Casts are the large, transparent, and waxy-looking objects which are occasionally seen, and are thought by some to denote a corresponding condition of the kidney. But nothing definite is yet known concerning them. They sometimes present a few transverse markings and fissures, as though they were very brittle and had been broken.

This completes the list of casts, and it may be useful to sum up their clinical significance.

The appearance of a few small hyaline casts may be, and probably is, the result of a congestion in the kidney of recent origin.

Epithelial and hyaline casts are very frequently found in the urine of patients recovering from scarlet fever, pneumonia, bronchitis, and congestive diseases generally.

Fatty, granular, and waxy casts have a grave significance; yet the appearance of one or two of them should not induce us to pronounce too certainly the fatal termination of the case.

It is the continuance of a prevailing kind of cast which is most to be relied upon in the way of diagnosis and prognosis.

DETECTION.—We depend upon the microscope altogether for the detection of casts, and considerable skill is sometimes necessary to distinguish between them and certain extraneous matters. The most frequent sources of confusion in this respect are cotton fibres and feathers. Cotton fibres are striated, and are apt to be folded or twisted. A particle of feather or hair has definite anatomical characters, which should be too well known to allow of a mistake. The appearances of these accidental substances are shown in Figure 10.

Fig. 10.

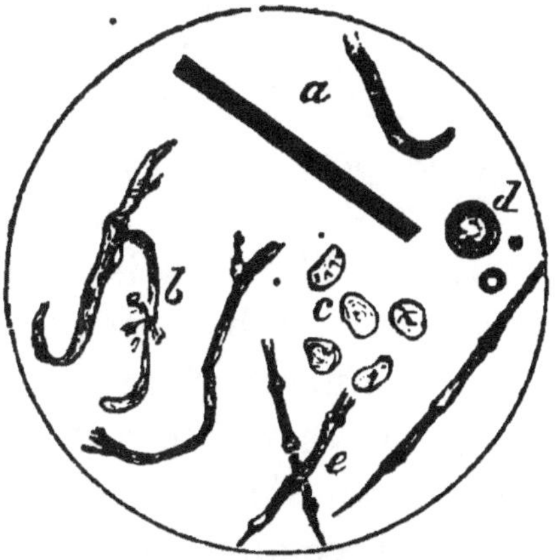

a, Hairs; *b*, Cotton fibres; *c*, Starch grains; *d*, Air-bubbles; *e*, Feathers.

As a guide, it may be said of casts, that they are generally rounded at one extremity, have parallel sides, and are not flexible enough to allow of folding or twisting.

Casts are liable to disintegrate, and may become so changed as to escape detection if allowed to remain in decomposing urine. Therefore it is important to look for them before decomposition begins. The lightest cast will have fallen to the bottom of the urine-glass in ten or twelve hours. Then gently pour away all the urine except about two drachms, and from this, with the aid of a glass tube, take up, from the bottom, a drop or two for microscopic examination. (See last page for detailed directions for microscopic manipulations.)

Pus.—There are various causes giving rise to pus

in the urine. Cystitis, pyelitis, gonorrhœa—in fact, an abscess or ulceration communicating with the urinary tract at any point. In women, purulent discharges from the vagina are likely to confuse us as to the origin of the pus found in their urine.

Urine containing pus is turbid when voided, and, on standing, deposits a whitish cloud of a ropy consistency, which distinguishes it from an inorganic deposit. It decomposes very readily and emits a nauseous odor.

Tests.—Pus consists of a fluid and a corpuscular element. The fluid is albuminous, and will be acted on accordingly by heat and nitric acid, although purulent urine never gives a marked coagulum; and this fact serves as a good test, the turbidity not disappearing with heat and nitric acid, and at the same time scarcely becoming more marked.

Liquor potassa causes a semi-solid, gelatinous precipitate.

The microscope will reveal the pus corpuscles, and this is conclusive evidence. These bodies are a little larger than the blood corpuscles, colorless and spheroidal. They are made up of cell-wall, granular contents, and nuclei. By the addition of a drop of acetic acid, the cell-wall is rendered transparent, and the nucleus is brought sharply into view. (Fig. 13.)

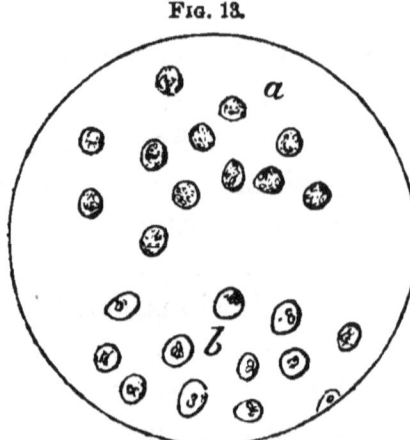

Fig. 13.

a, Pus corpuscles; *b*, Effect of acetic acid.

ALALYSIS OF THE URINE. 55

Oil.—It has been asserted that oil is a constituent of normal urine in very small proportion, and only detected after a careful analysis. It, however, sometimes makes its appearance in quantities sufficient not only to be recognized under the microscope, but visible to the unassisted eye.

It may present itself either as distinct oil-drops, as granules, or as a very fine emulsion, each particle of which appears as a mere point under the microscope (this latter condition we have in *chylous urine*). If either of these forms of fat are in the urine, we shall have a layer of it on the surface when allowed to rest. In the case of granular fat, the layer will appear creamy; where the globules are of any size, a yellow, oily layer will be seen.

Oil globules were detected in the urine of a man who was taking cod-liver oil by Roberts. Fatty degeneration of the kidneys is accompanied by oil in the urine. Much speculation has been indulged in concerning the source of oil where there is no recognizable affection of any organ. It is known that during digestion, the blood is loaded with chyle, and may, under circumstances not understood, permit it to escape into the urine. (See "Chylous Urine.")

A frequent source of oil-drops in the urine is the vessel in which it is collected, or the bottle in which it is brought to the physician, and, in many instances, the passage of catheters and sounds, which are always oiled before being introduced.

Tests.—Oil is soluble in ether. But if it be in small quantity, it often becomes necessary to first extract it with ether, and afterward evaporate almost to dryness before any traces of it can be seen.

Under the microscope, oil-drops are distinguished by their perfectly circular outline, difference in size, and solubility in ether. As seen in urine, the globules are never very large, and might be mistaken for blood corpuscles. But their sharp-cut, bright outline, absence of color, and peculiar properties mentioned above, can scarcely fail to identify them.

Chylous Urine.—Chylous urine has ever been one of the most interesting, and at the same time puzzling conditions with which we have to deal. In appearance, it is milky, and on standing collects in a creamy layer on the surface. There is always more or less blood, fibrine, and albumen present. Sometimes it coagulates spontaneously when passed, and very closely resembles blanc mange. Cases are related where it coagulated in the bladder and completely blocked up the urethra, from which it was extracted in long flakes. It is a rare affection, and but few cases are reported. Writers disagree widely concerning its pathology and symptoms; but without going over the history at length, the following are the main points regarding this peculiar disease:

It is most common in warm climates; makes its appearance suddenly, and as suddenly ceases, to reappear again after months or even years. Sometimes it coagulates spontaneously, like lymph, and again does not undergo this change. The milkiness is more marked after meals. The older authors considered the kidneys and the assimilative functions of the system to be at fault and diseased. But casts have been searched for in vain, and several post-mortem examinations of individuals who were affected with this disease have failed to afford evidence of alterations in any organ. So that now it begins to be stated that the chyle and lymph are discharged directly into the urinary passage from the lymphatic vessels themselves; and Roberts especially advances this opinion, having noticed, in patients voiding chylous urine, appearances which indicated disease of the lymphatics.

We had an opportunity recently of examining a specimen of chylous urine, the history of which it may be well to relate. The patient had resided in the South most of his life. About ten weeks before, he had attempted to pass his urine five times within an hour. This necessitated great straining, and he was suddenly alarmed by a severe pain and a discharge of blood and milky urine. (The pain was located in the prostatic urethra, and as he had undergone an operation for stone a few years before, it and the hemorrhage were referred to that cause.) This condition came and disappeared several times until we saw the urine. The patient stated that the milky fluid sometimes was perfectly free from any urine. He was able to know this from the fact that the uriniferous odor was entirely absent, and, moreover, the bladder had just been emptied only a few minutes before.

This urine did not coagulate spontaneously. It contained blood, and had an alkaline reaction.

TESTS.—Besides the characters above stated, other tests are scarcely necessary. Chyle is oil in a state of emulsion.

Under the microscope, the granules appear very minute.

Spermatozoa.—The spermatic elements sometimes become mingled with the urine in sufficient numbers to form a deposit. As a deposit, they resemble mucus and pus, though never so abundant. Their presence may be accounted for as a result of coition or an involuntary discharge of semen. This latter may be continuous and constitute spermatorrhœa. With the exception of this last-named condition, the presence of spermatozoa in the urine is without significance.

Tests.—The semen is albuminous, and will be rendered cloudy, to a more or less degree, by heat and nitric acid. But the microscope will decide whether the albuminous reaction is due to the presence of semen, in that it will reveal the characteristic filaments or spermatozoa. If the specimen be recent, they will be in active motion. Spermatozoa are possessed of a head and tail-like extremity, the former being slightly flattened from before backward. Their length is about $\frac{1}{600}$ of an inch. Fig. 14, a.)

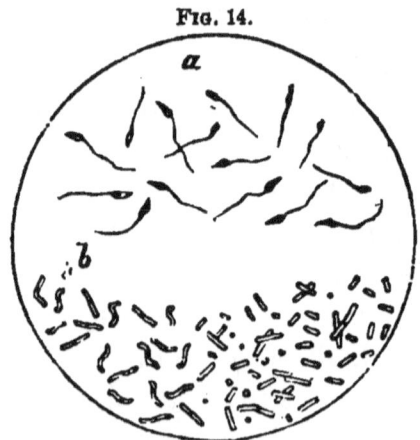

Fig. 14.
a, Spermatozoa; b, Bacteria.

Cystine.—This is a substance but rarely met with, compared with other deposits. Its source is not positively known, though supposed to be the liver. A remarkable fact concerning it is its liability to run in families. The most important clinical significance attaching to it is its liability to form a calculus; otherwise cystine may appear in the urine of an individual for years and not depreciate the health.

Urine containing cystine has usually an oily appearance, and deposits a light, rose-colored powder. Decomposition takes place very soon, and according to Dr. Bird, the color changes to a green. Cystine contains sulphur, which in the process of decomposition is evolved as sulphuretted hydrogen.

TESTS.—Acetic acid will cause a further precipitate to take place. It is insoluble in the vegetable acids, and is not dissolved by heat. It is soluble in ammonia and mineral acids.

The crystals are six-sided plates, and colorless. If a solution of cystine be placed in a shallow dish with a little ammonia, evaporation will cause a deposit of the pure crystals. Cystine crystals are easily distinguished from uric acid by their solubility in nitric acid. (Fig. 15.)

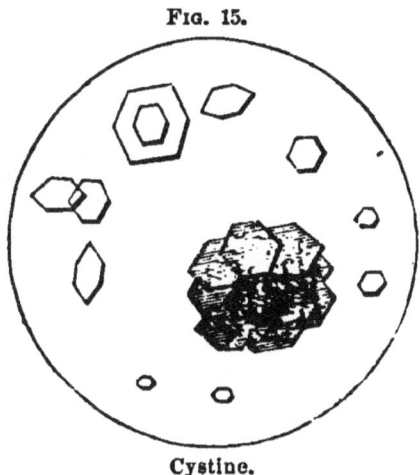

Fig. 15.

Cystine.

Keistine.—It was at one time thought that the urine of pregnant women offered peculiarities by which the pregnant state could be diagnosed. It was stated that there would form upon the surface of such urine a layer of cheesy matter unlike any appearance presented when the subject was not pregnant.

The name keistine (cheesy) was given to this formation. The results of observation, however, have been so conflicting that the profession, to-day, do not regard the formation as a distinct one, but attribute the phenomenon as one and the same thing as the alkaline fermentation, which is hastened in pregnancy by the presence of an increased amount of animal matter, such as epithelium and mucus from the vagina and bladder.

VEGETABLE FUNGI.

There are several microscopic vegetable growths which invade the urine. The most important are the *Penicillium glaucum*, or common mould, and *Saccharomyces*, or sugar fungus.

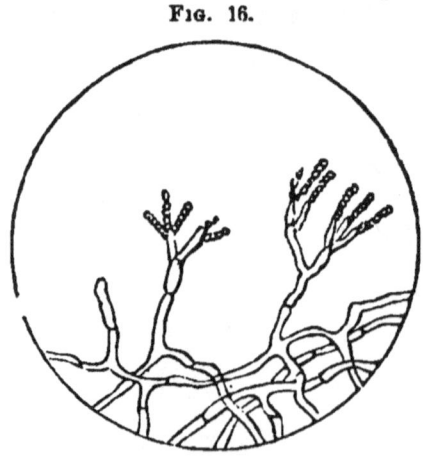

Fig. 16.

Penicillium Glaucum.

Penicillium Glaucum.—This is the common mould so frequently seen on old leather and articles kept in damp places. It is apt to appear in urine when left exposed at ordinary temperature, but whether there will be a visible deposit depends upon the extent of the growth.

This fungus as seen under the microscope consists of two distinct parts: the mycelium and the spores.

The mycelium or vegetative portion is an irregular interlacement of fibres which are immersed in the urine. From these arise slender stalks which shoot upwards towards the surface, and on reaching it develop the spores upon their extremities.

It is chiefly by the arrangement of the spores upon the stalk that we are able to distinguish the different fungi having a mycelium. In the variety under consideration they are disposed in diverging rows and have a peculiar whitish-green color; hence its name. (*Penicilus*, brush-like, and γλαυκος, sea-green.)

The spores ripen and fall into the urine or are blown away to other quarters. In either case, if the locality be a favorable one, they immediately begin to elongate and then to branch, until very soon a mycelium of new growth is formed.

There are other varieties of vegetable fungi very similar to penicillium liable to be encountered in a microscopic inspection of urine, but none of them have any special interest, their presence being due to the ubiquitous nature of their spores.

Saccharomyces.—This is the term applied to the variety of fungus found in solutions undergoing the vinous fermentation—conversion of sugar into alcohol and carbonic acid. It will make its appearance in any ·saccharine mixture containing albuminous matter and subjected to a temperature of about 70° F. Saccharomyces is therefore especially interesting to the physician in that its appearance in urine is a certain indication of the presence of sugar.

This plant consists simply of cells of about 10 micro-millimetres ($\frac{1}{8500}$ of an inch) in diameter, of an oval or spheroidal shape, generally colorless and having a granular appearance, with one or more germinal bodies in their interior. There are several species, differing chiefly in the size and shape of the cells; that of beer-yeast is known as "Saccharomyces Cerevisiæ." Saccharomyces multiplies by budding, and during the active state the cells present appearances indicative of the process; from the young bud just beginning to emerge from the parent cell, on through the various stages of its growth until it has attained an equal size, and finally separates. Occasionally the new cell puts forth buds before it becomes independent and gives rise in this way to a chain or irregular collection of cells, as seen in the figure.[1]

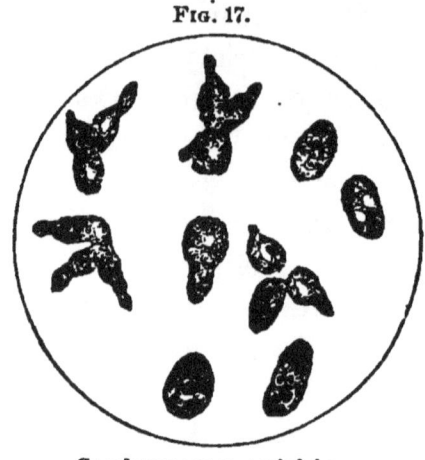

Fig. 17.

Saccharomyces cerivisiæ.

Bacteria (little rods).—This is the general term given to the minute vegetable organisms invariably present in putrefying animal and vegetable matter. They consist of simple cells filled with a colorless fluid and presenting several varieties of form: spheroidal (micrococcus), curved (vibrio), twisted (spirillum), and oblong, (bacterium).

The varieties most common to decomposing urine are bacterium termo, vibrio, and micrococcus; but unless we employ very high powers of the microscope

---

[1] For an elaborate description of these and similar fungi, see Botanische untersuchungen über die Alkoholgärungspilze. By Max. Reess, Leipsic. 1870.

these last will not be detected, on account of their very minute size.

Bacterium termo measures about $\frac{1}{500}$ mm. ($\frac{1}{12800}$ of an inch) in length by about $\frac{1}{1000}$ mm. ($\frac{1}{24000}$ of an inch) in breadth. Vibriones are of about the same width, but three or four times longer.

The most remarkable feature connected with these bodies is their active motion, which is seen to consist of rapid vibrations with the effect to propel the cell with considerable velocity through the fluid.

The appearances presented by bacteria of two cells connected together is due to their mode of multiplication being by division; if closely watched the separation will very soon be seen to be complete.

PART IV.

ACCIDENTAL INGREDIENTS WHICH DO NOT FORM DEPOSITS.

THERE are certain abnormal substances frequently present in urine, but which are in such perfect solution as to afford no evidence of their presence by way of a deposit, and with the exception of *bile* may have little or no effect upon the natural color.

Such are *albumen*, *sugar*, and *bile*.

Albuminuria.—We have seen that urine containing pus, blood, or semen will necessarily be albuminous, but we come now to consider albuminuria which is not due to the presence of either of the above but which is occasioned by the escape of the albuminous constituents of the blood through the kidneys. This phenomenon may result from

1. Mal-assimilation of albuminous food.
2. Mechanical obstruction to the renal circulation.
3. Chronic degeneration of the kidneys.

We know that in order to be assimilated albuminous substances must first be converted into albuminose, and subsequently into blood-albumen. Now it is a well established fact that if this process is not carried out and albumen enters the circulation imperfectly digested, it is immediately eliminated by the kidneys. It is therefore easy to understand how deranged digestion or excessive indulgence in albuminous food will be sometimes followed by albuminuria.

Mechanical obstruction will include all of those conditions which interfere with the normal circulation of the kidneys and induce congestion. The cause may be an inflammation of the kidney itself as in nephritis (Bright's disease); it may be due to passive congestion from hepatic or cardiac disease or the pressure of a tumor, and it is induced by the specific poisons of many general affections, as rheumatism, gout, scarlet, yellow, typhus and typhoid fevers, and malaria. Increased blood-pressure is also brought about in the kidneys by lesions of the central nervous system. In these the vaso-motor filaments are paralyzed.

Chronic degenerations of the kidneys are, as a rule, the result of prolonged congestions; and the albumen which escapes them is principally due to a continuance of this condition and the frequent attacks of inflammation to which they are especially liable.

Albuminous urine has certain characteristics appreciable by simple ocular inspection. In the first place it froths more than normal urine, when violently agitated; and if the disease be of long standing the color will be light and the specimen present a peculiar hazy appearance due to the presence of casts, fat granules and epithelium. In nephritis there will be more or less blood present.

The specific gravity of albuminous urine is low—1004–1015—on account of the deficiency in solids, especially urea.

It is of the highest importance to subject urine in which albumen has been detected, to a microscopic examination; for it is by the presence or absence of casts and certain other debris that we are able to judge of the morbid conditions which prevail. (See "Casts.")

Albumen always exists in urine where casts are found, but casts are not necessarily present in every instance with albumen.

TESTS.—Albumen is coagulated by heat, alcohol, the mineral acids and by a solution of acetic acid and potassium ferrocyanide. Other methods do not concern us.

A test-tube is one-third filled with the suspected urine, and held in the flame of a spirit-lamp until boiling is fairly going on. If any perceptible opacity has ensued it can be due to two things—albumen or the earthy phosphates. The question is determined by the addition of one or two drops of nitric acid; if it be albumen the opacity is slightly increased, if the phosphates it entirely disappears.

When the amount of albumen present is very small it is necessary to proceed very carefully and have an equal quantity of the same specimen, which has not been manipulated, in another test-tube for comparison. Nitric acid alone is a delicate test for albumen. The test-tube should be inclined and the acid allowed to run down the side and gain the bottom, where it will be seen as a clear layer. If albumen be present there will appear an opaque stratum where the urine and acid come in contact.

There are two fallacious results which are likely to follow the addition of nitric acid to urine containing an excess of either urea or the urates, which might mislead a novice. With the first there may be a crystalline precipitate of urea nitrate; with the second an amorphous deposit of uric acid. Neither of these, however, should be taken for albumen, for they both disappear when heated; and besides the urea salt presents definite crystals as will the uric acid if allowed to stand a few hours.

The most delicate test for albumen is with acetic acid and potassium ferrocyanide. Render the fluid to be tested distinctly acid with acetic acid and add a few drops of a solution of potassium ferrocyanide; the presence of albumen will be indicated by an increased opacity.

There is one fact to be borne in mind in testing with heat alone where no evident effect follows its application. It is that a strongly alkaline condition will hold albumen in solution against heat; hence the importance of always employing nitric acid after boiling.

A convenient, and for general purposes a sufficiently accurate, method for estimating the quantity of albumen from day to day, is simply to coagulate with heat and nitric acid and allow the coagulum to collect at the bottom of the test-tube and compare its mass with the amount of urine used, as $\frac{1}{4}$, $\frac{1}{2}$, etc. For greater accuracy a graduated tube may be used.

Sugar.—There is a grave disorder chiefly characterized by the excretion of enormous quantities of saccharine urine of high specific gravity, 1030–1060, known as diabetes mellitus. The pathology is obscure. We know that one of the functions of the liver is to produce sugar, and that this process is carried on independent of saccharine and starchy food. But what is the particular lesion or class of lesions which causes the fluids of the body to be saturated with sugar is not determined.

It has been suggested that diabetes mellitus may be due to two distinct conditions:—An excessive production of sugar by the liver, and a failure of the system to assimilate that normally produced.

Experimentally sugar can be made to appear in the urine by various operations upon the nervous system and by any means which accelerates the portal circulation.

The ingestion of large quantities of sugar or food containing it, especially after fasting, will induce a saccharine condition of the urine. Disease of the liver is not, then, necessarily the primary cause of diabetes.

Sugar appears in the urine, temporarily, during the course of various affections, as,—disease and injury of the nervous system, mental shock, amputation, pregnancy, etc.

Saccharine urine presents a peculiar bright straw color. Albumen is frequently present. A drop placed upon the fingers and allowed to evaporate leaves a sticky molasses-like residue.

TESTS.—If saccharine urine be boiled with liquor potassæ, it will assume a dark brown or molasses color. This is known as *Moore's Test*. It is unreliable in that any urine of high specific gravity will give the same result.

If a little yeast be added to urine containing sugar, *fermentation* will ensue, whereby carbonic acid gas and alcohol are produced. The gas escaping in bubbles can be collected and tested. This is a good test, but not a very convenient one. It is not necessary to add yeast except to hasten the result, for if sugar is present in urine the yeast fungus, saccharomyces cerevisiæ, will be spontaneously developed and the decomposition proceed as in the former case. A convenient apparatus with which to apply the fermentation test to urine is constructed as follows: Procure a glass jar or test-tube capable of standing upright, and provide it with a tight-fitting stopper. Through the stopper let a glass tube descend almost to the bottom of the jar. Half an inch of the lower extremity of this tube should be bent upwards, and that portion which is outside the jar curved so as to have its extremity directed downwards. Fill the jar with the suspected urine, add a small lump of German yeast, and adjust the cork and tube. The whole is now to be kept in a moderately warm place, 70° F., for about 48 hours; and if sugar be present, except in very minute amount, the decomposition into alcohol and carbonic acid gas will take place. The gas accumulating at the top of the jar and unable to escape, presses upon the urine below, and drives it out of the long glass tube. Having collected a quantity of gas, the cork is carefully removed and the usual tests for carbonic acid applied.

Trommer's Test.—This method of determining the presence of sugar in a solution is based upon the fact that sugar possesses the property of immediately reducing the salts of copper in an alkaline solution at the boiling-point.

The usual mode of applying this test is to add two or three drops of a solution of copper sulphate to a quantity of the suspected fluid in a test-tube, and then pour in an excess of liquor potassæ. If sugar be present, the whole now assumes a deep, transparent blue color.[1] Upon boiling, copper oxide will be thrown down as a bright yellow precipitate (in some cases an orange red). When this change occurs, the blue color disappears entirely, and the mixture becomes perfectly opaque.

Unfortunately when applied to the urine after the above manner, these phenomena do not present themselves. Having added the copper and potassa, there will appear the usual transparent blue color, but when boiled we almost invariably fail to see anything like a yellow or red precipitate. Either the blue color is entirely destroyed, and a dark, transparent, molasses color appears, or there is a dirty green precipitate.

Considerable has been written concerning the inapplicability of Trommer's test to the urine, the explanation being that the organic constituents, urea, coloring matter, etc., interfere with the reduction of the suboxide of copper. Directions have therefore been given to get rid of these matters by filtration through finely-powdered animal charcoal. This process is an efficient, but as a general rule, not a convenient one for the practicing physician.

[1] If a solution of copper sulphate be treated with potassium hydrate there will be a pale blue precipitate of copper hydrate formed ; but if we add a small quantity of tartaric acid previous to the potassium the precipitate will be redissolved on the addition of an excess of the latter.

Sugar has a property similar to tartaric acid in this respect, and the very fact that on the addition of the reagents of Trommer's test to a fluid results in the production of a perfectly clear blue color, is in favor of the presence of sugar. We should never be satisfied however with this evidence alone, especially in dealing with the urine.

Therefore, because I believe this method to be the most simple and reliable one, I submit the following hints and rules for the application of Trommer's test to the urine; they being the result of a series of investigations on the subject.[1]

In the first place, *it is necessary to have a great excess of the test,* because if *the urine be in too great quantity, the precipitate is dissolved;* and on this fact depends my method.

Proceed as follows: Place about 4 c.c. (1 ʒ.) of the urine in a test-tube, and to it add about 4 drops of a copper sulphate solution made in the proportion of 1 part of copper sulphate to 8 parts of water. To this mixture add liquor potassæ until the milky precipitate first formed is dissolved and the whole assumes a perfectly transparent blue color.

The conditions now are such that the precipitate of copper oxide will be in such amount as not to be totally dissolved by the small quantity of urine employed. Therefore when this blue mixture is boiled the reaction is as satisfactory as with a watery solution of sugar. If there is no immediate reaction set the test-tube aside and wait ten or twenty minutes. There is no change if the urine is free from sugar.

If albumen is present the color resulting from the mixture of the test fluids will assume a purplish hue, and may interfere with the process. It can be removed by coagulating and filtering.

Do not mistake the flocculi of the phosphates, thrown down by the alkaline test fluid, for a precipitate of the suboxide of copper. This latter is dark

[1] New York Medical Journal. June, 1874, p. 632.

red or yellow, and soon subsides to the bottom of the test-tube, and is seen there as a compact little mass.

The test with FEHLING'S SOLUTION (for composition, see p. 75) is based upon the same principle as the foregoing, and if we have the solution, it is a very easy and reliable test. But it has to be prepared, and when kept for any length of time, is liable to undergo changes which unfit it for further use.

The way to use it is to take a test-tube a quarter full and boil it. To this add a drop of the suspected urine if much sugar is present; if little, add ten or fifteen drops.

Here we have the same principle as advised with Trommer's test—namely, a small quantity of urine and a great excess of the test fluid.

If any change occurs in the Fehling liquor when boiled alone, it is unfit for use.

Bile.—The coloring matter of the bile is frequently excreted with the urine, imparting to it a more or less greenish-brown color.

It occurs in jaundice even before the skin has become perceptibly colored, and continues a little while after the natural color is restored.

The biliary salts, glykocholate and taurocholate of soda, sometimes are present also, and it is frequently important to know whether the bile is represented in the urine solely by its coloring matter, or whether these more important ingredients are there too.

TESTS.—The coloring matter of bile can be detected by pouring a little of the urine into a white plate, and allowing it to come in contact with a few drops of nitric acid. As the two mingle, a play of colors will be observed, varying from a violet to a green. Any oxidizing agent, tincture of iodine, or the atmosphere, will produce a grass-green color.

Instances may occur when the bile pigment is present in such small amount as to render its detection difficult. In such cases the urine should be allowed to remain at rest several hours at a low temperature to favor the deposition of the urates. If these constituents are precipitated they will appropriate the bile color and by separating them and dissolving in a small amount of water with heat, we can then apply the nitric acid test as above.

To detect the biliary salts we must resort to "*Pettenkofer's test,*" as follows: If the color of the urine be very marked it should be mixed with animal charcoal and filtered; then take about 10 c.c. (℥ ii) of the clear liquid thus obtained and to it add a few drops of cane-sugar (one part to four of water). Sulphuric acid is now to be added very cautiously, the test-tube being occasionally dipped in cold water to keep the temperature thus developed at about 60°. When the acid is first added if the biliary salts are present there will be a whitish precipitate of cholic acid, which subsequently disappears on the continued addition of the reagent. The next change consists in a cherry-red color which appears at the bottom of the tube. Cease, now, to add the acid and observe that the red gradually changes to a deep purple.

It is this play of colors which constitutes the most characteristic feature of the test.

If the urine contains albumen it should be coagulated and removed, as this substance will imitate in some degree the behavior of the biliary matters with Pettenkofer's test.[1]

[1] See Dalton's Physiology, 1875, p. 212, for other precautions to be observed with Pettenkofer's test.

PART V.

QUANTITATIVE ANALYSIS.

SUGAR—UREA.

SUGAR and urea are the only substances of which it is frequently of practical importance to estimate the amount in urine.

Sugar.—There are two methods by which the physician can readily determine the quantity of sugar in a given solution; one necessitating considerable skill in chemical manipulation, and the other remarkable for its simplicity.

The first of these is based upon the reactions with Trommer's test, the only conditions being that we must use a graduated solution, it having been found that a certain amount of sugar will reduce a definite quantity of copper sulphate.

The solution commonly employed is known as "Fehling's liquor." This investigator, having discovered that 5 parts of sugar decompose 34.64 parts of copper sulphate, advised the following solution:

Pure crystallized copper sulphate	. .	34.64 grammes.
Neutral potassium tartrate	. . .	150 "
Solution of sodium hydrate (sp. gr. 1.12)		650 "

The potassium tartrate is dissolved in the solution

of caustic soda, and the copper sulphate in a little pure water. The two solutions are now mixed and the whole diluted with water until the volume amounts to 1 litre. 10 c.c. of this solution will be decolorized by .05 grammes of sugar. The objection to Fehling's solution thus prepared is its liability to spontaneous decomposition when kept for any length of time, some of the copper being reduced by the tartaric acid which is set free. The following method is therefore generally employed and is much preferable:

1st. Pure dry crystallized copper sulphate, 34.64 grammes. Distilled water, 1000 grammes. 10 c.c. of this solution correspond to .05 gramme of sugar.

2d. Solution of potassium hydrate (sp. gr. about 1050.

3d. A saturated solution of potassium bitartrate.

These three solutions are to be kept in separate glass-stoppered bottles, and are not mixed until used. They therefore do not deteriorate by keeping. The apparatus necessary for this analysis consists of a glass chemical flask of 250 c.c. (8 oz.) capacity; burette graduated to cubic centimetres and tenths; spirit lamp and stand upon which to place the flask to boil its contents.

The process is as follows: 10 c.c. of the copper sulphate solution are first put into the flask and diluted with about four times as much water. About 10 c.c. of the potassium tartrate solution are now added and well mixed, after which pour in the solution of potassium hydrate until the whole assumes a clear blue color.

Our test-fluid is now ready. Set it to boil. Then take 2 c.c. of urine and dilute it with 18 c.c. of water;[1] the mixture now consists of $\frac{1}{10}$ urine. By this time ebullition has commenced in the flask and we proceed to add, drop by drop, the mixture of urine and water and closely watch the result. As the saccharine urine comes in contact with the boiling copper solution some of the latter is decomposed and appears as a red precipitate. A time will at length arrive when the whole mixture changes to a brick-dust red, and no blue color remains. The entire operation should cease every little while to allow the red copper oxide to settle, when it will be easy to determine whether any blue color remains in the supernatent liquid.

During the last part of the process great care is necessary in order not to add any more urine after the blue color has disappeared; if we do not stop at the right time, the mixture will suddenly assume a molasses color, the result of the action of the boiling alkali upon the sugar, and our analysis is worthless. By a little practice proper skill is readily acquired.

Suppose then the blue color has entirely disappeared from the mixture and we find that 30 c.c. of the diluted urine have been used; but this only contains $\frac{1}{10}$ or 2 c.c. of urine. Therefore 2 c.c. of the urine contain .05 gramme of sugar. The percentage is easily calculated with this system of weights and measures:

$$2 : .05 :: 100 : x = 2.5 \text{ per cent.}$$

Perhaps the best method to determine the quantity

[1] This is in order to add the saccharine urine more gradually. If the sugar is not abundant this dilution is not necessary.

of sugar in urine, adapted to the busy physician, is that recommended by Roberts,[1] which is based upon the difference in specific gravity before and after fermentation.

The specific gravity of the specimen is taken and written down; a bottle provided with a perforated cork is then about half-filled, and a lump of German yeast added, about the size of a walnut, and the whole placed where it will remain at about 80° F. for eighteen hours. At the end of this time the decomposition into alcohol and carbonic gas will be complete; and now every degree of density lost indicates 2.188 milligrammes of sugar per cubic centimetre before fermentation, (one grain for every fluid ounce).

Urea.—The daily quantity of urea excreted frequently becomes of importance, and various methods of estimating it have been advised. None are so simple as the one proposed by Dr. Davy, of England.[2] Doubts having arisen as to its accuracy, we undertook a series of experiments in order to ascertain if they were well-founded. The results of these experiments[3] seemed to show that the method is in every respect sufficiently accurate for practical purposes, and it is therefore introduced here as the one best adapted to the physician who wishes to perform the analysis for himself.

This method depends upon the decomposition which ensues when urea is brought into contact with sodium, potassium, or calcium hypochlorite.

Nitrogen gas is evolved, and being collected and

[1] Urinary and Renal Diseases, p. 198. Philadelphia, 1872.
[2] Philosophical Mag. 1854.
[3] New York Med. Jour., Sept. 1872.

ANALYSIS OF THE URINE. 79

measured, the amount of urea originally present is estimated. The following are Dr. Davy's directions:

"A strong glass tube, about twelve or fourteen inches long, closed at one end, and its open extremity ground smooth, and having the bore not larger than the thumb can conveniently cover, holding from two to three cubic inches (50 c.c.), each divided into tenths and hundredths by graduation on the glass, is filled more than a third full of mercury, to which afterward a measured quantity of urine to be examined is poured, which may be from a quarter of a drachm to a drachm or upward, according to the capacity of the tube. Then, holding the tube in one hand, near its open extremity, and having the thumb in readiness to cover the aperture, the operator fills it completely full with a solution of sodium-hypochlorite (taking care not to overflow the tube), and then instantly covers the opening tightly with the thumb, and having rapidly inverted the tube once or twice, to mix the urine with the hypochlorite, he finally opens the tube under a saturated solution of common salt and water, contained in a steady cup or mortar. The mercury then flows out, and the solution of salt takes its place, and the mixture of urine and hypochlorite being lighter than the solution of salt, will remain in the upper part of the tube, and will therefore be prevented from descending and mixing with the fluid in the cup. A rapid disengagement of minute bubbles of gas soon takes place in the mixture in the upper part of the tube, and the gas is there retained and collected. The tube is then left in the upright position till there is no further appearance of minute globules of gas being formed, the time being dependent upon the strength of the hypochlorite and the quantity of urea present. But the decomposition is usually completed in from three to four hours; it may, however, be left much longer, even for a day if convenient, and having set the experiment going, it requires no further attention; and when the decomposition is completed, it is only necessary to read the quantity of gas produced off the scale on the tube. In cases where great accuracy is required, due attention must be paid to the temperature and atmospheric pressure, and certain corrections made if these should deviate from the usual standards of comparison, at the time of reading off the volume of gas; but in most cases, sufficiently near approximation to accuracy may be obtained without reference to those particulars."

It has been found by calculation that 1 cubic centimetre of gas corresponds to 2.5 milligrammes of urea —(1 cubic inch to .64 grains).

There are one or two sources of error to be avoided. Ammonia and uric acid will give rise to nitrogen and thereby increase the apparent amount of urea. It is the former which is most likely to exist in quantities sufficient to cause confusion; but as it only occurs in appreciable amount during the alkaline decomposition we can always avoid its effect by using fresh urine.

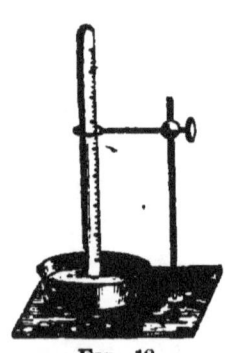

Fig. 18.

The apparatus necessary to perform this analysis is shown in Fig. 18. The graduated tube can be procured of Benjamin & Co., No. 10 Barclay street.

Sodium hypochlorite is preferable, to either that

of calcium or potassium because it is easily procured, being an article in general use under the name "Liquor Sodæ Chlorinatæ," or Labarraque's Solution. But there are various preparations of this solution, both foreign and domestic, which do not correspond in strength, and are not universally kept on sale. We are in the habit of employing a preparation of reliable manufacture, and one most generally sold in the United States, namely, "Squibb's Liquor Sodæ Chlorinatæ," and advise its use in connection with this analysis.

4*

PART VI.

CALCULI AND GRAVEL.

WE have seen that the urine is subject to a variety of deposits. Now, under certain conditions some of them are precipitated within the urinary passages in a manner to form solid masses of considerable size.

When the masses are sufficiently small to pass through the ureters or urethra they are called *gravel* and *sand;* when too large to admit of this passage they are designated as *calculi*.

It is important to be aware that the predisposing causes of the small concretions conduce alike to the formation of the larger.

Calculi.—By a urinary calculus or stone, we mean those solid concretions found in the bladder and kidney, and which vary in size from that of a pea to a hen's egg. They differ in composition as well as in general form and texture.

As regards composition, the most common are *uric acid, lime oxalate, urates, phosphates,* and *cystine*.

A stone is rarely composed of one substance alone; but is made up of alternating and varying concentric layers of different deposits.

In the first place the center of almost every urinary concretion consists of either uric acid or lime oxalate with a little mass of mucus; and the rest is composed of layers of the same, alternating occasionally with the earthy phosphates. And if the process is uninterrupted the stone will finally become encrusted with successive deposits of this last named material.

Calculi originate in various ways. A little mass of retained mucus or blood will induce the acid decomposition, and determine the formation of uric acid and lime oxalate crystals in the substance of the mass, and thus constitute the nucleus of a stone. Foreign substances as hair pins and pieces of slate pencil are not infrequently found to have originated a phosphatic stone. Finally, as already mentioned when studying deposits, we may have a spontaneous precipitation of crystaline or amorphous matter taking place within the body as a result of retention, and consequent decomposition, or on account of a derangement in the natural proportions of the urinary constituents.

Whatever be the origin, the subsequent history of calculi is soon told. There are alternating layers of various materials until the mass begins to act as an irritating foreign body. The bladder now becomes inflamed; retention and decomposition ensue, and thus the very presence of the stone conduces to its own rapid growth. Although we are not likely to meet with a stone composed entirely of one material, yet this may happen, or at any rate, one particular substance may predominate and impart its own characters to the mass. This is especially true when removal is effected early, prior to the alkaline decomposition.

Uric Acid.—Uric acid is the most common of all deposits going to form calculi. The nucleus of almost every stone is composed of crystals of this substance, and not unfrequently we see it constituting a stone almost entirely, especially if it be removed before the alkaline fermentation ensues.

When this is the case, the size rarely exceeds that of a pigeon's egg. The color ranges from a brick-dust red to a fawn, and the surface is generally covered with little tubercles; if, however, there were more than one stone in the bladder, their surfaces would be worn smooth by attrition.

They are usually spherical, and slightly flattened. When sawed in half, their concentric laminæ are very distinctly seen.

When a stone of this composition exists in the bladder, the urine is acid.

Tests.—The murexid test is the most direct and characteristic. Place a fragment of the stone in a clean porcelain capsule, and add one or two drops of nitric acid; now evaporate the acid by gentle heat over the flame of a spirit-lamp, and a pink color will appear, which changes to a purple when a little ammonia is dropped in.

Uric acid is soluble in an alkali from which it may be precipitated by hydrochloric acid and its characteristic crystals seen under the microscope.

Under the blow-pipe it is converted into a black or very dark-brown coal.

Urates.—Stones composed wholly of the urates are rare, though this deposit frequently alternates with others.

When pure, they are small, and are apt to originate in the kidneys. Most frequently these calculi are met with in children. These formations are of a red or brick-dust color. The urine will be acid.

TESTS.—The fact of their solubility in warm water will prove the presence of the urates.

Lime Oxalate.—Like most other deposits, the lime oxalate more frequently forms a stone by alternation than alone. When this substance exists alone or predominates, the stone is exceedingly hard and rough; in fact, it has been called the *mulberry calculus*, from its warty and irregular surface.

When the concretions of this formation are small enough to be classed as gravel, and are voided with the urine, they may be smooth. Here, however, there is a mixture of the urates, and the formation has probably taken place in the kidneys, and by their number the surfaces are worn.

In color, these stones are sometimes dark brown or even black, and again almost white.

Most mixed calculi contain layers of oxalate of lime.

The urine will be strongly acid.

TESTS.—Oxalate of lime is soluble in the mineral acids. Under the blow-pipe, it is reduced to a dark ash which turns moistened red litmus-paper blue,—caustic lime being the residue of the combustion.

PHOSPHATES.

We have seen that an alkaline condition of the urine will precipitate the earthy phosphates, and that the alkalinity may be due to two causes, only one of which is concerned in the formation of phosphatic calculi. We have said that the alkalinity of the fixed salts may continue for a length of time without a stone being found; because the phosphatic precipitate will then be amorphous, and has little tendency to form concretions. But when putrefaction occurs from retention of urine, or the action of inflammatory products anywhere in the urinary passages, then we have ammonium carbonate developed, and now *crystals* of the ammonio-magnesian or triple phosphate are thrown down, as well as the amorphous phosphate of lime.

It is the ammoniacal decomposition, then, which most influences the formation of phosphatic calculi.

Sometimes, however, after prolonged administration of alkaline medicines, and as a result of high living (indulgence in wine) where the urine is kept alkaline by the fixed salts, concretions of AMORPHOUS LIME PHOSPHATE may form.

These stones are light colored, easily broken, and present an earthy fracture. They rarely alternate with other forms.

TESTS.—Lime phosphate is insoluble in water; soluble in a weak acid, and is practically infusible under the blow-pipe.

Mixed Phosphates.—Deposits of this nature are the result of the ammoniacal decomposition of urine, and consist of the crystalline ammonio-magnesian phosphate with the amorphous lime phosphate. They rarely constitute the whole or interior of a stone, but are generally added as the result of the presence of a pre-existing calculus, the irritation of which has induced the decomposition just mentioned. And it is a fact, that almost every stone, if it remains long enough in the bladder, will be encrusted with several layers of the mixed phosphates. These calculi sometimes attain to enormous dimensions.'

They are chalk-colored, and have a glistening appearance on account of the presence of the crystals of the phosphates on their surface.

TESTS.—Under the blow-pipe, the mixed phosphates readily fuse into a white enamel (and they have therefore been sometimes called the *fusible calculi*), and give off ammonia and water. They are soluble in weak acids.

Cystine.—This substance sometimes forms calculi and gravel. The number of cases is very limited, and very few specimens of the concretion exist. The stone is quite small, not larger than an almond, and of a dirty brown or greenish color. There appear to be no distinct laminæ, but rather a radiating structure. The pathology is obscure.

TESTS.—Cystine is soluble in the mineral acids, and in caustic ammonia; insoluble in water, acetic acid, and ammonium carbonate. If a piece of the calculus is dissolved in a little caustic ammonia, contained in a shallow dish, upon evaporation the characteristic hexagonal crystals already described, can be detected under the microscope.

Cystine burns with a bluish flame, emitting a peculiar, unpleasant, acid odor.

SCHEME FOR EXAMINATION OF URINE.

Proceed in the following order :
1. Odor.
2. Color.
3. Transparency.
4. Reaction to test-paper.
5. Specific gravity.
6. Daily quantity.
7. Any deposit ? Character of.
8. Apply reagents.

	PRECIPITATES.	DISSOLVES.
Heat.	Albumen. Phosphates.	Urates.
Nitric Acid.	Albumen. Urea. Uric acid.	Phosphates.
Liq. Potassæ.	Phosphates.	Albumen.
Acetic acid.	Cystine.	Albumen.
Silver nitrate.	Chlorides.	
Barium Chloride.	Sulphates.	

There are certain precautions to be observed with these reactions, for which reference must be made to the text.

GENERAL DIRECTIONS.

Urine intended for examination should be placed in a conical-shaped glass vessel, such as an ordinary ale-glass, in order that any deposit which may appear shall be concentrated in a comparatively small space; for if there is a very minute quantity, we otherwise might not be able to secure a specimen for the microscope, or detect its presence by ordinary inspection.

After the specimen has stood for six or eight hours, very gently pour off all but about half an ounce. This is done in order to prevent the deposited crystals, casts, or whatever is there, from becoming again mingled with a great quantity of urine when it is agitated by our subsequent manipulations.

Now, having provided yourself with a drop-tube, which be sure is clean, place the finger over the large end, and direct the other end to the bottom of the glass; raise the finger, and the urine ascends in the tube; replace the finger, and we have it confined there. Touch the point of the tube thus supplied upon a clean glass slide, and a drop will escape, which is all-sufficient. A thin glass cover is placed over this drop on the slide in a manner to exclude air-bubbles, as follows: place one side of the cover on the slide and allow it to come gradually down. If the cover floats, there is too much urine, which will overflow it, and obscure the lens. Under such circumstances, just absorb the surplus urine with a little blotting-paper.

We need a good $\frac{1}{4}$-inch lens for the examination of urinary deposits, and sometimes a $\frac{1}{8}$-inch in order to distinctly discern small hyaline casts.

LIST OF APPARATUS AND REAGENTS REQUIRED.

APPARATUS.

Absolutely necessary.
$\begin{cases} \text{½ doz. 4-inch test-tubes.} \\ \text{Test tube rack.} \\ \text{Alcohol lamp.} \\ \text{Small glass funnel.} \\ \text{Filter-paper to fit funnel.} \\ \text{Red and blue litmus paper.} \\ \text{Urinometer.} \qquad \$2.00. \end{cases}$

Special apparatus.
$\begin{cases} \text{Glass flask 200.c.c. (8 ʒ).} \\ \text{Cubic centimetre measure.} \\ \text{Glass tube graduated to cubic centimetres.} \end{cases}$
$\begin{matrix} \text{Quantitative analysis of sugar.} \\ \\ \text{Quantitative analysis of urea.} \\ \$2.00. \end{matrix}$

REAGENTS, IN GLASS-STOPPERED BOTTLES.

Nitric acid, 4 ʒ.
Acetic acid, 4 ʒ.
Sulphuric acid (pure), 4 ʒ.
Solution caustic potassa (20 gr.–1 ʒ), 8 ʒ.
Sol. copper sulphate (1 ʒ–1 ʒ), 4 ʒ.
Sol. silver nitrate (10 gr.–1 ʒ), 4 ʒ.
Sol. barium chloride (10 gr.–1 ʒ), 4 ʒ.

$3.50.

COMPARATIVE VALUES OF FRENCH AND ENGLISH MEASURES USED IN THIS BOOK.

1 gramme = 15.434 grains.
1 cubic centimetre = 0.061 cubic inch = 0.27 ʒ.
1 litre = 61 cubic inches = 33.8 ʒ.
1 millimetre = $\frac{1}{25}$ inch.
1 micromillimetre = $\frac{1}{25000}$ inch.

INDEX.

A.

	PAGE
Acid, Effect of, upon Normal Urine	19
Acid Fermentation	21
Albumen	45, 64
Abnormal Urine	25
Alkalies, Effect of, upon Normal Urine	19
Alkaline Fermentation	22
Apparatus and Reagents required	90

B.

Bacteria	62
Barium Chloride, as Reagent	20
Bile	72
Blood	43
Black Urine	26
Blue "	26

C.

Calculi	82
" Cystine	87
" Phosphates	85
" Lime Oxalate	85
" Urates	84
" Uric Acid	83
Casts	47
" Blood	51
" Epithelium	50

	PAGE
Casts Fatty	50
" Granular	51
" Hyaline	49
" Pus	51
" Waxy	52
Changes in Urine on Standing	21
Chlorides	19
Chylous Urine	56
Color of Normal Urine	13
Color of Abnormal Urine	25
Cold, Effect of, upon Urine	18
Composition of Urine	5
Creatinine	13
Cystine	58, 87

D.

Daily Quantity of Normal Urine	17
" " " Abnormal Urine	29
Deposits	30, 43
Directions for examining Urine	89

E.

Effects of Reagents upon Normal Urine	18
Epithelium	39
" Bladder	40
" Renal	40
" Urethral	40
" Vaginal	40
Extraneous Matters in Urine	53

F.

Fehling's Solution, Composition of	75
" " Used as Qualitative Test	72
Fermentation Test for Sugar	68
Fungi, vegetable	60

G.

	PAGE
Gravel	82
Green Urine	26

H.

Hæmaturia	44
Hæmatinuria	44
Heat, Effect of, upon Normal Urine	19
Hippuric Acid	85

K.

Keistine	59

L.

Lead, Basic acetate of, as Reagent	20
Lime Oxalate	21, 38
" " Calculus	85

M.

Medicines, Reappearance of in Urine	20
Moore's Test for Sugar	68
Mould Fungus	60
Mucus	41

N.

Normal Urine, Characters of	5

O.

Odor of Abnormal Urine	25
Oil in Urine	55

P.

Penicilium Glaucum	60
Pigments	42

Phosphates.. 19, 22, 28, 36
" Ammonio-magnesian 22, 37, 86
" Earthy, Precipitated by Alkalies............ 19
" Stella....................................... 36
" Calculi... 85
Pus... 53

Q.

Quantitative Analysis...................................... 75
" " of Sugar............................. 75
" " of Urea.............................. 78

R.

Reaction of Normal Urine................................. 15
" " Abnormal Urine............................ 27
Reagents, Effects of, upon Normal Urine................... 18
" Required....................................... 90

S.

Saccharomyces Cerevisiæ................................. 61
Scheme for examining Urine............................. 88
Silver Nitrate, as Reagent............................... 19
Specific Gravity of Normal Urine......................... 16
" " " Abnormal Urine.... 29
Spermatozoa... 58
Sugar, Tests for................................... 67, 68
" Quantitative Analysis of......................... 75
" Fungus 61
Sulphates, Precipitation of............................... 20

T.

Transparency of Normal Urine............................ 15
" " Abnormal Urine........................... 27
Trommer's Test for Sugar............................... 69
" " adapted to Urine....................... 71

U.

Urates.. 18, 33
  " Calculi.. 84
Urea... 6
  " Nitrate.. 7, 8
  " Physiological and Pathological Relations of........... 9
  " Quantitative Analysis of............................. 78
Uric Acid... 19, 21, 30
  "   " Calculi... 83
Urinary Deposits.. 30

V.

Values, Comparative of French and English Measures....... 91
Vegetable Fungi.. 60
Vibriones.. 62

www.ingramcontent.com/pod-product-compliance
Lightning Source LLC
Chambersburg PA
CBHW020901160426
43192CB00007B/1023